Fanyu Meng

Modeling of Moving Sound Sources Based on Array Measurements

Logos Verlag Berlin GmbH

λογος

Aachener Beiträge zur Akustik

Editors:
Prof. Dr. rer. nat. Michael Vorländer
Prof. Dr.-Ing. Janina Fels
Institute of Technical Acoustics
RWTH Aachen University
52056 Aachen
www.akustik.rwth-aachen.de

Bibliographic information published by the Deutsche Nationalbibliothek

The Deutsche Nationalbibliothek lists this publication in the Deutsche Nationalbibliografie; detailed bibliographic data are available in the Internet at http://dnb.d-nb.de .

D 82 (Diss. RWTH Aachen University, 2018)

ISBN 978-3-8325-4759-2
ISSN 2512-6008
Vol. 29

Logos Verlag Berlin GmbH
Comeniushof, Gubener Str. 47,
D-10243 Berlin
Tel.: +49 (0)30 / 42 85 10 90
Fax: +49 (0)30 / 42 85 10 92
http://www.logos-verlag.de

Modeling of moving sound sources based on array measurements

Von der Fakultät für Elektrotechnik und Informationstechnik der
Rheinischen-Westfälischen Technischen Hochschule Aachen
zur Erlangung des akademischen Grades eines

DOKTORS DER INGENIEURWISSENSCHAFTEN

genehmigte Dissertation

vorgelegt von

M.Sc.

Fanyu Meng

aus Kedong, Heilongjiang, China

Berichter:

Univ.-Prof. Dr. rer. nat. Michael Vorländer
Univ.-Prof. Dr.-Ing. Peter Jax

Tag der mündlichen Prüfung: 05. Juli 2018

Diese Dissertation ist auf den Internetseiten der Hochschulbibliothek online verfügbar.

Abstract

When auralizing moving sound sources in Virtual Reality (VR) environments, the two main input parameters are the location and radiated signal of the source. An array measurement-based model is developed to characterize moving sound sources regarding the two parameters in this thesis. This model utilizes beamforming, i.e. delay and sum beamforming (DSB) and compressive beamforming (CB) to obtain the locations and signals of moving sound sources. A spiral and a pseudorandom microphone array are designed for DSB and CB, respectively, to yield good localization ability and meet the requirement of CB. The de-Dopplerization technique is incorporated in the time-domain DSB to address moving source problems. Time-domain transfer functions (TDTFs) are calculated in terms of the spatial locations within the steering window of the moving source. TDTFs then form the sensing matrix of CB, thus allowing CB to solve moving source problem. DSB and CB are further extended to localize moving sound sources, and the reconstructed signals from the beamforming outputs are investigated to obtain the source signals. Moreover, localization and signal reconstruction are evaluated through varying parameters in the beamforming procedures, i.e. steering position, steering window length and source speed for a moving periodic signal using DSB, and regularization parameter, signal to noise ratio (SNR), steering window length, source speed, array to source motion trajectory and mismatch for a moving engine signal using CB. The parameter studies show guidelines of parameter selection based on the given situations in this thesis for modeling moving source using beamforming. Both algorithms are able to reconstruct the moving signals in the given scenarios. Although CB outperforms DSB in terms of signal reconstruction under particular conditions, the localization abilities of the two algorithms are quite similar. The practicability of the model has been applied on pass-by measurements of a moving loudspeaker using the designed arrays, and the results can match the conclusions drawn above from simulations. Finally, a framework on how to apply the model for moving source auralization is proposed.

Contents

1 Introduction

Urban environmental noise has been receiving increasing attention by urban planners and decision-makers due to its impact on public health [1, 2]. Among the many sources of urban environmental noise, traffic noise caused by moving vehicles, such as cars and trains, is often the main contributor [3]. To sustain an acoustically comfortable urban environment, it is therefore essential to predict, assess and control traffic noise.

Sound pressure level (SPL) is the most commonly used metric to evaluate noise in urban spaces. SPL has the advantage of being a simple number. However, it ignores the human perception of sound, which is a significant factor when assessing soundscapes in urban environments [3]. Auralization, as an efficient technique compared to SPL, enables people to perceive simulated sounds intuitively, and allows non-acousticians to evaluate proposed acoustic scenarios and thus participate in the planning of the urban environment [4]. After decades of research and development in auralization, a good progress has been achieved particularly concerning powerful propagation simulation models and 3D audio technology. But significant challenges still remain, e.g. a lack of methods, data formats and standards for sound source characterization, which prohibits a vast extension of auralization into practice, although progress has been made concerning the human voice [4] and musical instruments [5].

Models used for sound synthesis according to their inherent principles can be classified by forward and backward models [6]. The forward model requires physical or spectral information, or relies on the generation mechanism of sound sources. In previous studies, prediction tools and empirical equations were used to generate the sound sources of aircraft [7, 8, 9, 10, 11, 12, 13, 14]. An emission synthesizer of a wind turbine was established by empirical equations [15]. Similarly, empirical equations were also applied on synthesizing the sound sources of an accelerating car [16], including tires [17]. Recently, physically based temporal synthesis models for rolling and impact noise of trains have been developed [18].

The backward model utilizes either near-field or far-field recordings to extract sound source signals. Compared to the forward model, undertaking measurements consumes much more time. Nevertheless, it saves the time to establish physical or empirical models. Moreover, synthesis from recordings overcomes the deficiency of low realism which is probably the main drawback of the forward model. Arntzen et al. [11] concluded that the synthesized sounds of aircraft flyovers were perceived differently compared to the measurements due to the empirical source models.

Similar results were reported when auralization was implemented in form of a web-based virtual reality (VR) tool, which synthesized signals based on forward modelling, resulting in artificial perception [19]. More problematically, theoretical models or empirical equations are not always achievable.

In [20], sound samples were extracted by analyzing near-field recordings of an engine running at various speeds. Target sounds were synthesized by concatenating corresponding samples with an overlap and add algorithm in real time. Additionally, in previous work, the synthesized signals were compared and validated with the original signals [6]. Validation using the backward model is more convenient since synthesized signals are obtained from measurements and thus comparison can be directly conducted after synthesis, whereas the forward model additionally needs specific validation measurements. A model for synthesizing electrical railbound vehicles was suggested by Klemenz [21]. In this model, rolling noise and air conditioner noise were added directly from recordings, while the tonal traction noise components were synthesized by simple calculations of sinusoids and sweeps.

If no near-field recordings are available, the backward model is needed to compute source signals from far-field recordings. For example, aerodynamic noise, caused by high-speed motion, is only possible to be measured at a comparatively large distance from the moving object. In this sense, the backward model can also be called inverse model. An aircraft auralization model was established using microphone recordings based on a backward sound propagation model [22, 23]. For an aircraft, the sound source can be considered as one point source due to the large measurement distance. However, for ground vehicles, e.g. cars and trains,

are typically placed closer to the measurement object, entailing that vehicles cannot be considered as single point sources anymore, but have to be extended to multiple independent point sources. Fig. 1.1 schematically depicts a pass-by measurement of a car with potentially relevant individual point sources recorded by microphones.

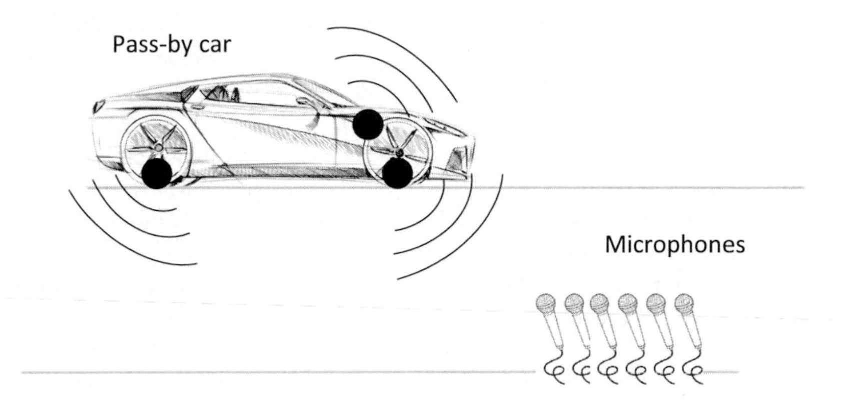

Figure 1.1.: A sketch of a pass-by car measurement with an array of microphones. The sound sources are represented by the solid dots.

The backward model was applied to synthesize sound sources of a train by back propagating mono recordings in terms of various locations as the positions of the sound sources [24]. Bongini et al. [25] addressed that sound sources should be represented by their locations, spectral signals and directivities during auralization. They used a two-dimensional microphone array to localize the sound sources on a pass-by train using beamforming, and then derived the signals and directivities of the sources by controlling the train passing slowly by its back-propagated signals recorded by a vertical array. Nevertheless, although the trains passed by slowly, the recording for a target source was still contaminated by other sources as in [25]. Besides, the positions of sound sources on a moving vehicle are mostly unknown.

To summarize, no proper general models for generating the source signals of moving sources, e.g. cars and trains for the purpose of auralization are available. Sounds generated by the forward model require a priori knowledge about the generation mechanism of the sources, and normally lack high fidelity. In addition, extra experiments for validation is another drawback. The backward model is able to overcome these shortcomings, however, the synthesized sources of ground vehicles delivered by the existing backward models are contaminated by other non-target sources. Therefore, further development of the backward model is desired.

Beamforming, as mentioned above [25], is a common post-processing algorithm based on microphone array data and has been widely applied for sound source localization [26, 27, 28]. Researchers have also been studying the localization for moving sources in the past decades [29, 30, 31, 32, 33] by eliminating the Doppler effect [34, 35, 36]. In addition, Sijtsma et al. [37] introduced the time-domain transfer function (TDTF) incorporated with Doppler effect to enable the localization of moving sources with arbitrary motions, for example, acceleration and circular motions [37]. However, DSB fails to yield high spatial resolution [38], which might result in reconstructed signals containing unwanted noise from neighboring sources. In order to reconstruct signals more precisely, beamforming methods with increased spatial resolution, which can be modified for moving sound sources are necessary. Higher spatial resolution can be achieved, e.g. by minimum variance distortionless response (MVDR) [39], multiple signal classification (MUSIC) [40], minimum power distortionless response (MPDR) and linear constrained minimum variance (LCMV) [27]. Super-resolution even in the presence of noise and reverberation is possible by means of the sparse recovery (SR) algorithm [41, 42] and cross pattern coherence (CroPaC) algorithm [43], or through compressive beamforming (CB) [44, 45, 46, 47, 48] using only a small number of microphones.

All methods mentioned above have been used with varying degree of success to localize sound sources, being it stationary or not. However, not all these methods are able to reconstruct the source signal. CB, as originally proposed, was utilized for the localization of moving sources [44, 47], however, not for the extraction of the source signal. Edelmann and Gaumond [46] mentioned the possibility to “listen to” the source by applying an inverse Fourier transform on the CB output, but it was not executed and yet the target was still on stationary sources. Therefore, DSB and CB will be explored with the focus on reconstructing non-stationary signals, in this way extending the application of beamforming algorithms for auralization.

In this thesis, the backward model using beamforming, i.e. DSB and CB, is applied for localization and extended for signal reconstruction for the purpose of auralizing moving sound sources.

This thesis therefore focuses on the following main contents:

- Microphone array design for DSB and CB;

- Extending DSB for the signal reconstruction of moving sound sources;
- Developing CB for the localization and signal reconstruction of moving sound sources;
- Guidelines for using the model are provided through parameter studies;
- Framework development of the array measurement-based model for auralization.

The outline of the thesis is as follows. Chapter 2 introduces the theoretical aspects of moving sound sources, DSB and CB, spectral analysis and synthesis, as well as the evaluation criteria of source localization and signal reconstruction. In Chapter 3, fundamentals of microphone arrays are introduced, including the design procedures of a spiral array for DSB, and a pseudorandom array for CB are demonstrated. To compare with DSB, designing the pseudorandom array also takes into account the localization performance of DSB. Chapter 4 extends DSB for the signal reconstruction of a periodic signal. The model using DSB is further evaluated with varying parameters, including steering window, window length and source speed. Pass-by measurements are performed to apply and validate the DSB model. Chapter 5 further develops CB for localizing and reconstructing a moving engine signal, in which CB and DSB are compared. The performance under various regularization parameters, window lengths, signal-to-noise ratios (SNRs), basis mismatches and distances between the array and source trajectory are investigated. Pass-by measurements are performed again to apply and validate the developed CB model for localization and signal reconstruction of moving sources. Chapter 6 proposes a framework of applying the array measurement-based model for auralizing moving sound sources in VR. Last but not the least, Chapter 7 concludes the thesis and provides an outlook for future work.

2

Fundamentals

This chapter introduces the fundamental theories used in the thesis. The sound fields generated by stationary and moving sound sources are first described. Subsequently, modified DSB and CB for moving sound sources are delivered. Moreover, time-frequency analysis and spectral modeling synthesis (SMS) are introduced for the purpose of transforming the time-domain beamforming outputs into the frequency domain and then synthesize. The evaluation criteria in terms of localization and signal reconstruction are finally provided.

2.1. Moving sound source

This section will introduce the sound radiation from a point sound source with stationary and moving status respectively, and the concept of de-Dopplerization, the technique to eliminate Doppler shift in the received signal.

2.1.1. Sound field generated by a stationary source

Point sound sources and spherical wave propagation are assumed and this assumption holds for all the following contents. If the positions of a source and a microphone are $\vec{x}_s$ and $\vec{x}_r$, respectively, the distance between the source and microphone is

$$R = \|\vec{x}_s - \vec{x}_r\|_2. \tag{2.1}$$

The signal radiated from the source and measured by the microphone is [49]

$$p(t) = \frac{\rho}{4\pi R} q'(t - \frac{R}{c}), \tag{2.2}$$

where ρ is the density of the air, $q'(t)$ is the first derivative of the volume velocity $q(t)$, and c is the speed of sound. $\rho q(t)$ is the source strength. With defining $s(t)$ as a characteristic function of the source which is equivalent to $\rho q'(t)$,

$$s(t) \equiv \rho q'(t), \tag{2.3}$$

Eq. 2.2 can be rewritten in the form of

$$p(t) = \frac{s(t - \frac{R}{c})}{4\pi R}. \tag{2.4}$$

Here, signal $s(t)$ is the signal which plays a main role in the following contents. It is the signal with the strength and characteristics of the source which can represent the sound source. Thus, $s(t)$ is the signal to be reconstructed.

For M_p microphones ($M_p \in \mathbb{Z}^+$), with referring to Eq. 2.4, the received signal by the mth microphone is expressed as

$$p_m(t) = \frac{s(t - \frac{R_m}{c})}{4\pi R_m}, \tag{2.5}$$

where $m = 1, 2, ..., M_p$. In the frequency domain, Eq. 2.5 can be transformed to

$$P_m(\omega) = \frac{1}{4\pi R_m} S(\omega) e^{-j\omega R_m / c}, \tag{2.6}$$

where P_m and S_m are the Fourier transform of $p(t)$ and $s(t)$, respectively. To simplify the notation, ω/c is replaced by k, thus kR is discussed in the following contents. The complex term incorporates the spatial characteristics of the microphones and is referred to as manifold vector [27]. If N focus points are scanned as potential sound sources, the manifold vector can be written in a matrix

$$\begin{bmatrix} \frac{e^{-jk(\|R_{11}\|)}}{R_{11}} & \frac{e^{-jk(\|R_{21}\|)}}{R_{21}} & \cdots & \frac{e^{-jk(\|R_{N1}\|)}}{R_{N1}} \\ \frac{e^{-jk(\|R_{12}\|)}}{R_{12}} & \frac{e^{-jk(\|R_{22}\|)}}{R_{22}} & \cdots & \frac{e^{-jk(\|R_{N2}\|)}}{R_{N2}} \\ \vdots & \vdots & \ddots & \vdots \\ \frac{e^{-jk(\|R_{1M_p}\|)}}{R_{1M_p}} & \frac{e^{-jk(\|R_{2M_p}\|)}}{R_{2M_p}} & \cdots & \frac{e^{-jk(\|R_{NM_p}\|)}}{R_{NM_p}} \end{bmatrix}, \tag{2.7}$$

where $n = 1, 2, ..., N$.

2.1.2. Sound field generated by a moving source

When a sound source is moving following the trajectory of $\vec{x}_s(t)$, the distance between the source and the stationary microphone $\vec{x}_r$ is

$$R(t) = \|\vec{x}_s(t) - \vec{x}_r\|_2. \tag{2.8}$$

The sound pressure field produced by the moving source [50] is described by the following equation [1]:

$$\begin{aligned} p(t) =& \frac{\rho}{4\pi R(t)(1 - M\cos\theta(t))^2} q'(t - \frac{R(t)}{c}) \\ &+ \frac{\rho(v(cos\theta(t) - M))}{4\pi R(t)^2(1 - Mcos(\theta(t)))^3} q(t - \frac{R(t)}{c}), \end{aligned} \tag{2.9}$$

where $p(t)$ is the sound pressure at the microphone in the sound field generated by the moving source, $q(t)$ is the volume velocity of the source, $R(t)$ is the distance between the source and the microphone, v is the speed of the source, $M = v/c$ is the Mach number, and $\theta(t)$ is the angle between the moving direction of the source and source-microphone direction. A simple illustration of a point source moving rectilinearly at a constant speed of v is given in Fig. 2.1. Since the source is moving, R and θ are functions of time t. When the speed of the sound source is not too large compared to the sound speed, which is the case for current vehicles (high-speed trains can reach 300 km/h ($0.24M$)), and $R(t)$ is larger than 1 m,

[1] The expression of the original equation, Equation 11.2.15 in P724 in [50] is not consistent with the derivations in the preceding pages. Eq. 2.9 in this thesis gives the corrected expression.

the second term in Eq. 2.9 can be neglected. Again, this equation is rewritten using the equivalent signal $s(t)$ and denoted as

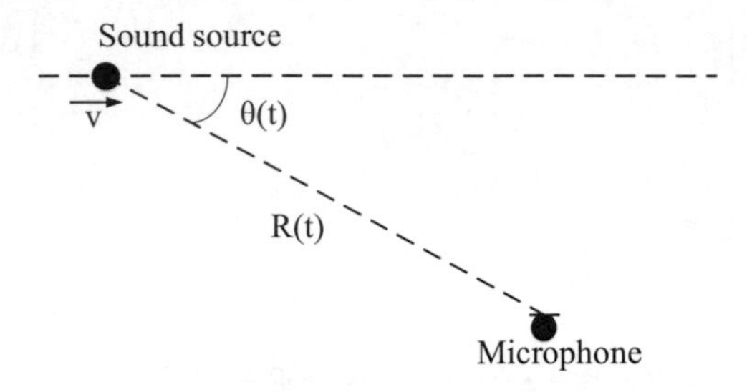

Figure 2.1.: Illustration of the rectilinear motion of a point sound source moving at a constant speed.

$$p(t) = \frac{1}{4\pi R(t)(1 - M\cos\theta(t))^2} s(t - \frac{R(t)}{c}). \tag{2.10}$$

In the digital world, no continuous-time signal is achievable. Thus the recordings of moving sound sources with microphones are discretized and simulated virtually. Appendix A shows the simulation of microphone recordings and the MATLAB codes used in this thesis.

2.1.3. De-Dopplerization

De-Dopplerization is the procedure to eliminate the Doppler effect [51]. In order to implement de-Dopplerization, the microphone array is supposed to track the moving sound source by steering the beam angle with respect to the spatial positions of the source along the moving trajectory. In this way, there is no relative movement between the microphone and the source, and thus the microphone is assumed to be moving with the source simultaneously. The following part will elaborate the de-Dopplerization procedure.

When a microphone is used to record a moving source, the recorded signal is uniformly sampled with equal time intervals. If the emission time t_e at the source is taken as the reference time during the following calculation and consider t_e as uniformly sampled, the reception time t at the microphone can be calculated by

$t = t_e + \frac{R_e(t_e)}{c}$, leading to non-uniformly spaced reception time samples t due to the non-linearity of $R_e(t_e)$. Moreover, $R_e(t_e)$ and $R(t)$ are distinguished because they are calculated in terms of different time variables t_e and t, knowing that $R_e(t_e) = R(t)$.

Therefore, first, the sampled microphone signal is interpolated in terms of the calculated non-uniformly spaced time stamps and is denoted as $\tilde{p}(t) \approx p(t_e + \frac{R_e(t_e)}{c})$ (Fig. 2.2(a)). According to Eq. 2.10,

$$\hat{s}(t_e) = 4\pi R_e(t_e)(1 - M\cos\theta_e(t_e))^2 \tilde{p}(t_e + \frac{R_e(t_e)}{c}), \tag{2.11}$$

where $R_e(t_e)$ and $\theta_e(t_e)$ are calculated in terms of t_e, and $\hat{s}(t_e)$ is the estimated source signal (Shown in Fig. 2.2(b)). Furthermore, defining the distance between the source and the "moving" microphone as R^0, the de-Dopplerized signal can be denoted as (Fig. 2.2(c))

$$\hat{p}(t_e) = \frac{1}{4\pi}\frac{\hat{s}(t_e - \frac{R^0}{c})}{R^0}, \tag{2.12}$$

or

$$\hat{p}(t) = \frac{1}{4\pi}\frac{\hat{s}(t - \frac{R^0}{c})}{R^0}. \tag{2.13}$$

In this thesis, $R^0 = min(R(t))$, where the source is closest to the microphone. Fig. 2.3 indicates how R^0 is defined.

2.1.4. Transfer functions for moving sound sources

Since $t = t_e + \frac{R_e(t_e)}{c}$, Eq. (2.10) can be written as

$$p(t) = \frac{1}{4\pi R_e(t_e)(1 - M\cos\theta_e(t_e))^2} s(t_e). \tag{2.14}$$

The time-domain transfer function (TDTF) is denoted by [52, 37]

$$H(t_e) = \frac{1}{4\pi R_e(t_e)(1 - M\cos\theta_e(t_e))^2}, \tag{2.15}$$

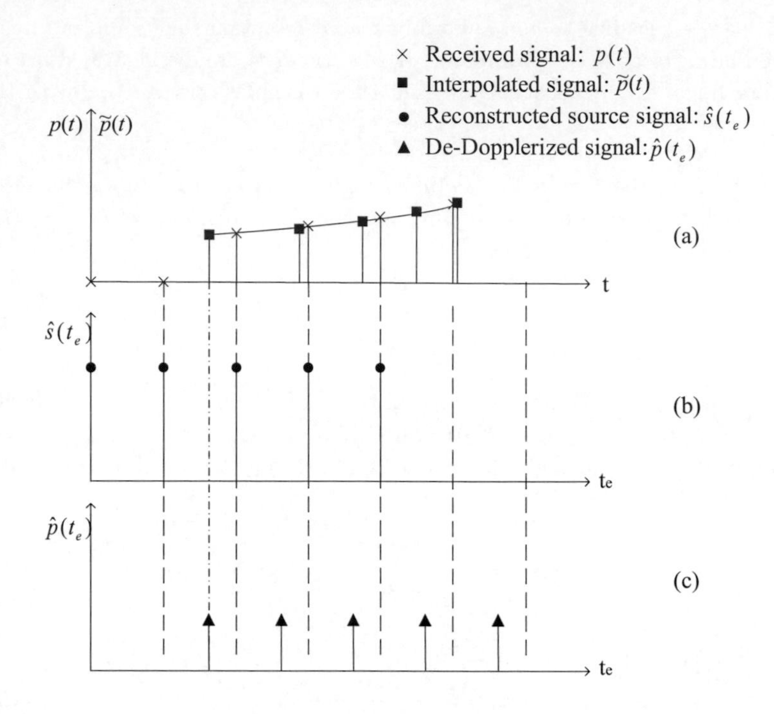

Figure 2.2.: De-Dopplerization procedure. a). The received signal (with the "cross" symbol) and the interpolated signal (with the "square" symbol); b). The source signal is reconstructed according to Eq. 2.11; c). The de-Dopplerized signal at the "moving" microphone calculated by Eq. 2.13.

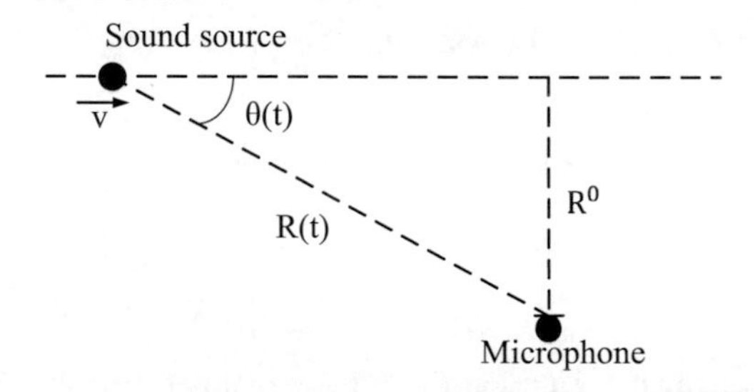

Figure 2.3.: The closest distance between the microphone and the source motion trajectory $min(R(t))$ is selected as R^0.

and it leads to

$$p(t) = H(t_e)s(t_e). \tag{2.16}$$

Here, the distance $R(t)$ is calculated by the emission time t_e at the microphone. In practice, noise is present in any measurement model. Therefore, Eq. (2.16) with additive white Gaussian noise is denoted by

$$p(t) = H(t_e)s(t_e) + n(t). \tag{2.17}$$

The time index t and t_e will be suppressed to simplify the notation. The problem above is extended to M_p microphones and N potential sources, which yields the following form

$$\mathbf{p} = \mathbf{Hs} + \mathbf{n}, \tag{2.18}$$

where $\mathbf{p} = [p_1, p_2, ..., p_{M_p}]^T$, $\mathbf{s} = [s_1, s_2, ..., p_N]^T$ represents the focus points of potential sources, $\mathbf{n} = [n_1, ..., n_{M_p}]^T$, and $H \in \mathbb{R}^{M_p \times N}$.

2.2. Beamforming

As mentioned in the Chapter. 1, beamforming has been a general way to localize sound sources based on temporal and spatial filtering using a microphone array [26, 28, 27]. This thesis is dedicated to address the localization and signal reconstruction of moving sources. Thus the beamforming algorithms are extended to moving sources, and their signal reconstruction problems. DSB and CB are the focused beamforming algorithms in the current work.

2.2.1. Delay and sum beamforming

DSB for a stationary source

The array is steered to different focus points to search for potential sound sources. For one focus point, the output of DSB is denoted as

$$y(t) = \sum_{m=1}^{M_p} w_m p_m(t + \tau_m), m = 1, 2, ..., M_p, \tag{2.19}$$

where $y(t)$ is the beamformer's output signal, M_p is the number of microphones, $p_m(t)$ is the signal received by the mth microphone, w_m is its amplitude weight, $\tau_m = (\hat{R}_m - \hat{R}_0)/c$ is the compensation of the time delay for the mth microphone in terms of the array origin, $\hat{R}_m$ is the distance between the focus point of the array and the mth microphone, and $\hat{R}_0$ is the distance between the focus point and the array origin. Uniform weight is used here, $w_m = 1/M_p$.

Now combining Eq. 2.5 and Eq. 2.19, the DSB output signal is expressed as

$$\begin{aligned} y(t) &= \sum_{m=1}^{M_p} \frac{w_m}{4\pi R_m} s(t - \frac{R_m}{c} + \tau_m) \\ &= \sum_{m=1}^{M_p} \frac{w_m}{4\pi R_m} s(t - \frac{\hat{R}_0}{c} + \frac{\hat{R}_m - R_m}{c}), m = 1, 2, ..., M_p. \end{aligned} \tag{2.20}$$

When the focus point coincides with the sound source position, $\hat{R}_m = R_m$ and $\hat{R}_0 = R_0$, Eq. 2.20 becomes

$$\begin{aligned} y(t) &= \sum_{m=1}^{M_p} \frac{w_m}{4\pi R_m} s(t - \frac{R_0}{c}) \\ &= \frac{1}{4\pi} (\sum_{m=1}^{M_p} \frac{w_m}{R_m}) s(t - \frac{R_0}{c}) \\ &= C_1 s(t - \frac{R_0}{c}), m = 1, 2, ..., M_p, \end{aligned} \tag{2.21}$$

where $C_1 = \sum_{m=1}^{M_p} w_m/4\pi R_m$ is a constant which depends on the weight and the positions of the potential sound source and microphones. The source signal $s(t)$ can be calculated by proper time shift and amplitude modification on the beamforming output signal $y(t)$. If $\hat{R}_m \neq R_m$, a degraded version of $s(t)$ would occur [28] and it leads to wrong localization, and thus wrong signal reconstruction.

DSB for a moving source

The de-Dopplerized signals can then be applied to DSB. Combining Eq. 2.19 and Eq. 2.13, the modified equation of DSB for a moving sound source is denoted as

$$\begin{aligned} y_{MS}(t) &= \sum_{m=1}^{M_p} w_m \hat{p}_m(t + \tau'_m) \\ &= \sum_{m=1}^{M_p} w_m \frac{1}{4\pi R^0_m} \hat{s}(t - \frac{R^0_m}{c} + \tau'_m). \end{aligned} \tag{2.22}$$

Here, $\tau'_m = (\hat{R}^0_m - \hat{R}^0)/c$, where $\hat{R}^0_m$ is the distance between the mth "moving" microphone and the focus point, and $\hat{R}^0$ is the distance between the "moving" array origin and the focus point. Similarly, if the focus point coincides with the source position, $\hat{R}^0_m = R^0_m$ and $\hat{R}^0 = R^0$, Eq. 2.22 becomes

$$\begin{aligned} y_{MS}(t) &= \frac{1}{4\pi}(\sum_{m=1}^{M_p} \frac{w_m}{R^0_m})\hat{s}(t - \frac{R^0}{c}) \\ &= C_2 \hat{s}(t - \frac{R^0}{c}), \end{aligned} \tag{2.23}$$

where $C_2 = \frac{1}{4\pi}\sum_{m=1}^{M_p}(w_m/R^0_m)$ is a constant which depends on the weight (here the weight is $1/M_p$) and the positions of the sound source and microphones. Similar to the stationary source, the estimated source signal $\hat{s}(t)$ can be reconstructed by time shift R^0/c and division of the constant C_2 on the beamforming output signal $y_{MS}(t)$.

2.2.2. Compressive beamforming

The concept of compressive sensing (CS) has been an emerging approach in image and audio processing. It asserts that a signal can be reconstructed with fewer measurements than conventional methods constricted by the Shannon Theorem [53]. This work aims at using CB for localization, and investigating the

CB output to reconstruct the signal of the moving source.

Description

$X \in \mathbb{C}^{N\times 1}$ represents the potential sound sources to be reconstructed. If X is sparse, it can be expressed in a sparse basis as $X = \Psi s (s \in \mathbb{C}^{M\times 1})$, where Ψ is the sparse basis matrix or sparse transform matrix that transforms signals from non-sparse basis to sparse basis. s is the sparse parameter, which has k nonzero values. In the absence of noise, the relationship between s and y is

$$y = \Phi X = \Phi \Psi s = As, \tag{2.24}$$

where Φ is the measurement matrix, and $A \in \mathbb{C}^{M\times N}$ is the sensing matrix, which is the product of the transform matrix Ψ and Φ. Actually, X and s represent the same signals but at different basis. When $M < N$, this is a underdetermined problem which has no unique solution. The method for solving this underdetermined problem is to converge all possible solutions to obtain the optimal solution according to the sparsity [47]. CS relies on two requirements: (1) sparsity of the signals and (2) sufficient incoherence of the mapping procedure from the source signals to the measurements [53]. In the spatial domain, sparsity implies that the number of sound sources is less than the number of focus points for array to scan. Incoherence expresses the idea that that objects having a sparse representation in Ψ must be spread out in the domain in which they are acquired, just as a Dirac or a spike in the time domain is spread out in the frequency domain [53].

A metric to measure the coherence of A [47] and CB relies on [54] is the Restricted Isometry Property (RIP). It is defined as: for each integer $p = 1, 2, \dots,$ the isometry constant δ_p of a matrix Φ as the smallest number to meet the requirement

$$(1 - \delta_p) \left\| s \right\|_2^2 \leq \left\| As \right\|_2^2 \leq (1 + \delta_p) \left\| s \right\|_2^2 , \tag{2.25}$$

for all p-sparse vectors. p-sparse means that a vector has maximum p nonzero entries [54]. The isometry constant δ_p of matrix A is as the smallest number. A satisfies the RIP of order p if $\delta_p \in (0, 1)$. The explanation of the RIP as the guarantee of incoherence is that all subsets of the p column(s) from A are nearly orthogonal. They cannot be exactly orthogonal since $N > M$. In this thesis, the 1-sparse case is adopted to test if the matrix A meets the RIP.

Solution

Usually the noise radiated by moving vehicles is generate by only a few sources, e.g. rolling, engine and aerodynamic noise for cars and trains [55, 56, 57, 18]. The presence of only a few sources enables exploiting the sparsity of $\mathbf{s}$ in Eq. (2.18). To impose the sparsity, the ℓ_0-norm problem needs to be solved. However, it is a non-convex problem which demands large computation time [47]. Fig. 2.4 shows the solutions of the ℓ_p-norm problems representing by the ℓ_p-balls with radius r, $\left\{s \mid \|s\|_p \leq r\right\}$ [47]. To compare with the ℓ_0- and ℓ_1-norm problems, the ℓ_2-norm is also given. Even the ℓ_2-norm problem is convex as the ℓ_1-norm, its aim is to minimize the energy of the signal through the ℓ_2-norm instead of its sparsity, leading to non-sparse solutions [47]. As the solutions shown in Fig. 2.4, only the ℓ_1-norm is equivalent to the ℓ_0 problem and is adopted to search the optimal sparse solution.

Previous studies applied the frequency-domain sensing matrix A to solve the problem [47, 48, 46], as well as in the time-domain [45] and the spatial domain [58]. However, they did not tackle the moving source problem. In this thesis, the sensing matrix is formed by the TDTFs, which incorporate the transfer functions in terms of the discrete spatial points of the source motion and thus incorporating the Doppler effect. In this way CB is extended to the moving source case.

Recalling Eq. 2.18, the ℓ_0-norm problem is

$$\min_{\mathbf{s}\in\mathbb{R}^N} ||\mathbf{s}||_0 \text{ subject to } \mathbf{p} = \mathbf{H}\mathbf{s} + \mathbf{n}. \tag{2.26}$$

Replacing it with the ℓ_1-norm

$$\min_{\mathbf{s}\in\mathbb{R}^N} ||\mathbf{s}||_1 \text{ subject to } \mathbf{p} = \mathbf{H}\mathbf{s} + \mathbf{n}, \tag{2.27}$$

which can be recast as the unconstrained optimization

$$\min_{\mathbf{s}\in\mathbb{R}^N} ||\mathbf{p} - \mathbf{H}\mathbf{s}||_2^2 + \lambda||\mathbf{s}||_1, \tag{2.28}$$

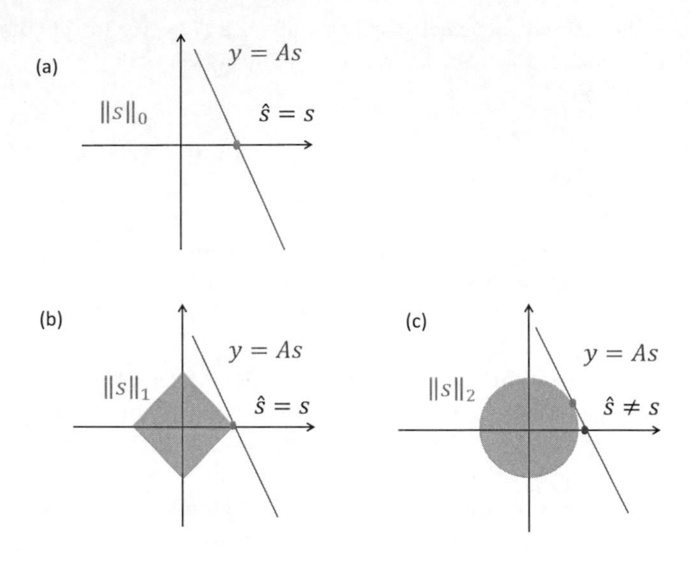

Figure 2.4.: Geometrical interpretation of the ℓ_0-, ℓ_1-, ℓ_2-norm problems.

where λ is the regularization parameter which balances the norm of the residual $||\mathbf{p} - \mathbf{H}\mathbf{s}||$ and the sparsity of $\mathbf{s}$ [44].

The stated solution is to solve the ℓ_1-norm optimization problem for a single time sample. For localization, single sample processing may find its application. However, if the values of one or some of the chosen time samples happen to equal or to be close to zero, the supposed spatial sparsity assumption would fail. In addition, signal reconstruction also requires more time samples to benefit the characteristic extraction in the frequency domain and signal synthesis. Therefore, the problem is extended to multiple time samples. The cost function is reformulated as

$$\mathbf{P} = \mathcal{H}\mathbf{S} + \mathbf{N}. \tag{2.29}$$

where $\mathbf{P} = [\mathbf{p}(t_1), ..., \mathbf{p}(t_T)] \in \mathbb{R}^{M_p \times T}$, $\mathbf{S} \in \mathbb{R}^{N \times T}$, $\mathcal{H} \in \mathbb{R}^{M_p \times N \times T}$ which samples $\mathbf{S}$ temporally and spatially, $\mathbf{N} \in \mathbb{R}^{N \times T}$ and T is the number of the time samples. Since sparsity is required in the spatial domain but not necessarily in time [59, 44], the ℓ_2-norm of all time samples of a focus point n is calculated, i.e. $s_n^{\ell_2} = ||s_n(t_1), ..., s_n(t_T)||_2$. With the ℓ_1-norm of $\mathbf{s}^{(\ell_2)} = [s_1^{(\ell_2)}, ..., s_N^{(\ell_2)}]$, the cost function becomes

$$\min_{\mathbf{s}^{\ell_2} \in \mathbb{R}^N} ||\mathbf{P} - \mathcal{H}\mathbf{S}||_F^2 + \lambda ||\mathbf{s}^{\ell_2}||_1, \tag{2.30}$$

where $|| \cdot ||$ represents the Frobenius norm. For a matrix $\mathbf{G} \in \mathbb{R}^{I \times J}$, $||\mathbf{G}||_F = \sqrt{\sum_{i=1}^{I} \sum_{j=1}^{J} |a_{ij}|^2}$.

As the reception time t is calculated from the emission time t_e, the interpolated recorded signal $\tilde{p}(t)$ is used as the substitution of $p(t)$ in Eq. (2.30) during the calculation. The ℓ_1-norm optimization problem is solved in MATLAB using the cvx toolbox [60]. After detecting the source position index n_s, the source signal reconstructed by CB is denoted as $\hat{s}_{n_s}(t), t = t_1, ..., t_T$.

To summarize, two beamforming algorithms, DSB and CB adapted for moving sound sources have been introduced. DSB and CB are both able to localize the spatial positions of the sound sources. More importantly, the source signals can be reconstructed from the beamforming outputs. Further information, e.g. spectral information from the reconstructed signals can be obtained for sound synthesis, in order to prepare for future auralization.

2.3. Time-frequency analysis

The beamforming algorithms are performed in the time domain. It is necessary to investigate the spectral information as well to further study the characteristics of the reconstructed signals in the frequency domain. This section introduces the common time-frequency transforms, i.e. Fourier transform (FT), short-time Fourier transform (STFT) and wavelet transform (WT).

2.3.1. Fourier transform

The Fourier transform (FT) is perhaps the most broadly used tool for the signal processing in science and engineering. Represented by a series of sinusoidal signals with various frequencies, a signal is transformed from the time domain to the frequency domain. The FT of a signal $x(t)$ with frequency f is denoted as

$$S(f) = \int_{-\infty}^{\infty} s(t)e^{-j2\pi ft}dt. \tag{2.31}$$

Although FT is a convenient time-frequency transformation, its limitation is also obvious. It delivers no time-variant frequency information since the integral in Eq.2.31 is conducted on the whole time axis.

2.3.2. Short-time Fourier transform

With the purpose of avoiding the loss of time information, Gabor proposed the window concept to slide the window uniformly along the time and perform FT respectively. This is the so-called Short-time Fourier transform (STFT), which overcomes the limitation of FT on the loss of time information after the time-frequency transformation [61]. By the inner product between the signal x(t) and the window function g(t), the STFT is defined as

$$STFT(\tau, f) = \int_{-\infty}^{\infty} x(t)g(t-\tau)e^{-j2\pi ft}dt, \tag{2.32}$$

where τ is the hop size of the window function. STFT delivers also the time information with the trade-off between time and frequency resolutions. When the window type is selected, the resolutions of time and frequency on the entire time and frequency scales are fixed.

STFT has not only the frequency but also the time information. It is beneficial to analyze the spectral characteristics of signals as it does not take the average over the whole time span. Each window can be regarded as a frame, where peak detection can be conducted, which will be introduced in Section. 2.4.

2.3.3. Wavelet transform

In recent years, wavelet analysis becomes popular especially in the field of image and signal processing in electric power systems since wavelet is able to provide multi-resolution analysis in the frequency domain without loss of the time information. Wavelet theory is based on the window function with variable length and originates from the research of Haar [20] in the beginning of the 20th century. By applying scaling s and time shift τ, the wavelet transform (WT) can be expressed as

$$WT(s,\tau) = \frac{1}{\sqrt{s}} \int_{-\infty}^{\infty} x(t)\psi^{\star}(\frac{t-\tau}{s})dt, \tag{2.33}$$

where $\psi(t)$ is the wavelet base. This expression is also called continuous wavelet transform (CWT).

WT is more flexible than STFT since the time and frequency scalings are adjustable. A finer time resolution can be obtained through WT. This benefits short-time signals, because if more frames are required, STFT is not sufficient to meet the requirement.

A comparison among FT, STFT and WT in reconstructing the frequency and amplitude of a harmonic signal has been deployed [62]. The results showed that using FT and STFT yielded more accurate reconstruction than using WT. Nevertheless, WT is promising if the parameters are properly selected.

2.4. Spectral modeling synthesis

Spectral modeling synthesis is a technique that analyzes and synthesizes signals separately with tonal and broadband components [63, 64]. The deterministic components consists of a series of sinusoids that can be represented by amplitudes and frequencies, and the broadband (stochastic) component is represented by spectral envelopes. In [63, 64], SMS is implemented through STFT. In this thesis, SMS is briefly introduced through CWT.

2.4.1. SMS Analysis

A signal is first segmented into frames. In each frame, CWT is implemented to transfer the signal in the frame from the time domain to the frequency domain.

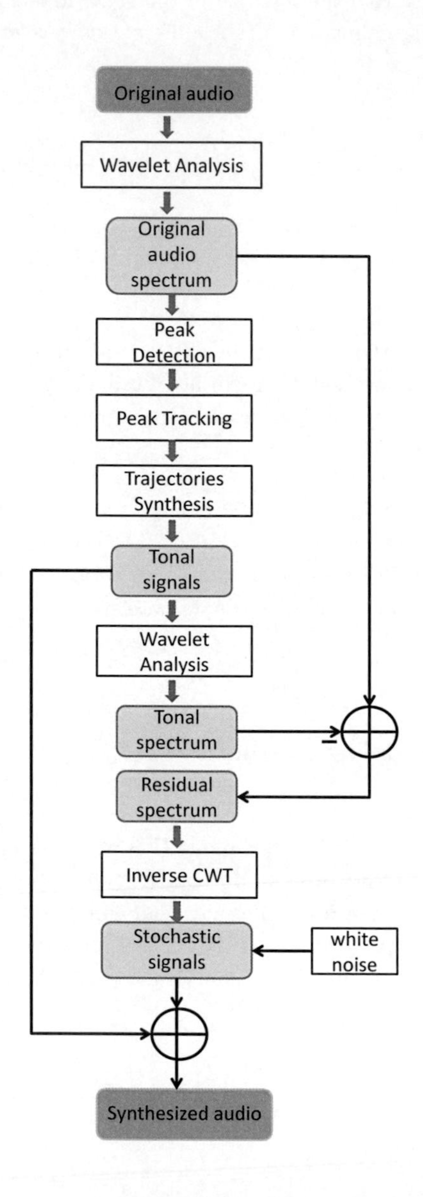

Figure 2.5.: Wavelet-based SMS diagram.

Peak detection

The center frequency F_c of CWT can be calculated by the function *centfrq* in MATLAB. If the wavelet is dilated by the scale a and the sampling frequency is fs, the value of the frequency F_a corresponding to the specific scale is defined as:

$$F_a = \frac{F_c \cdot f_s}{a}. \tag{2.34}$$

According to the Nyquist theorem, F_a varies from 0 to $f_s/2$. Therefore, the scale parameter varies from $2F_c$ to infinity. However, it is not possible to set an infinite value in the simulations. In practice, a scale which is large enough will be applied.

The wavelet spectrogram and the spectrum are plotted in Fig. 2.6. Fig. 2.6(b) is the spectrum of with the middle of the mother wavelet function overlapping with the first sample of the tonal signal. After obtaining the WT coefficients of all the frames, peak detection is conducted in each frame [64]. The found peaks in each frame are stored for further selection.

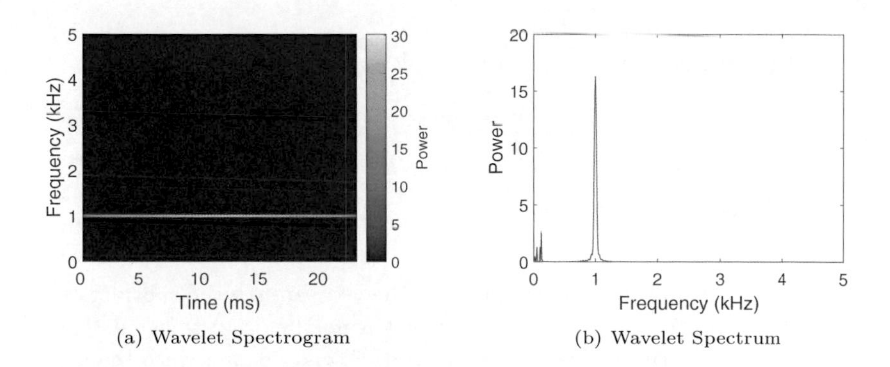

(a) Wavelet Spectrogram (b) Wavelet Spectrum

Figure 2.6.: Illustration of the spectrogram and spectrum of a tonal signal (0.023 s) of 1 kHz using WT.

Peak Continuation

After all peaks are detected for each frame, the peaks are tracked along the frames to check if every peak is continuous. The peak continuation algorithm is shown in Fig. 2.7 [63]. A particular tone can only be detected and saved in the end if it has been continuously detected, although disappearance in some frames is allowed. Otherwise, the peak is abandoned. The remaining peaks form trajectories along the frames. Each trajectory represents a tone that appears almost in the whole signal. The amplitudes and frequencies of the trajectories can be calculated consequently.

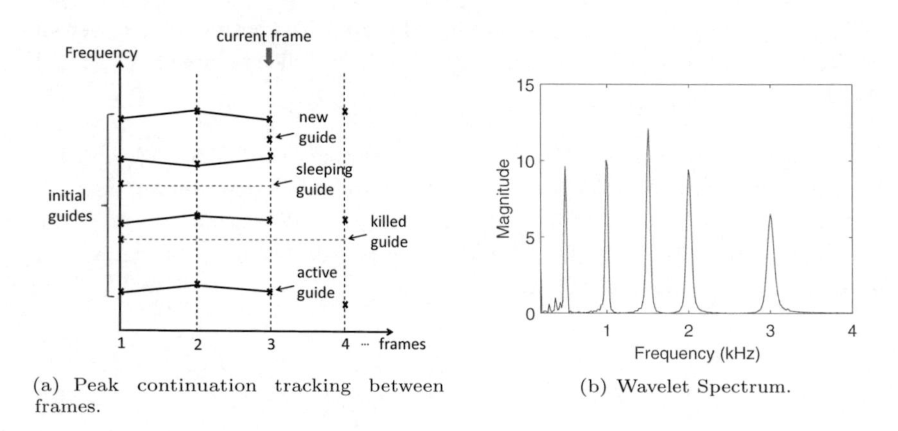

(a) Peak continuation tracking between frames.

(b) Wavelet Spectrum.

Figure 2.7.: Illustration of the peak continuation calculation.

Residual (Stochastic) component

The stochastic component is subsequently achieved through the subtraction of the detected deterministic (tonal) component from the original signal. It not recommended to perform the subtraction in the time domain since the phase information is not obtained from the peak detection and continuation procedure and thus it will lead to phase distortion [65]. Therefore, the subtraction is conducted in each frame in the frequency domain. For stochastic signals, the detailed magnitudes in the frequency domain are less perceptually influential than the envelope of the spectrum. The envelope is used for further synthesis.

2.4.2. SMS Synthesis

As both deterministic and stochastic components are calculated, the audio signal can be synthesized subsequently. Each sinusoidal component of the deterministic part can be synthesized separately and added together afterwards.

Tonal synthesis

Because of the lack of the phase information for each sinusoidal component, the deterministic part is synthesized using extracted frequency and amplitude information. They are added up in the following form:

$$S_d = \sum_{n=1}^{N} Acos(2\pi f_n t), \tag{2.35}$$

where f_n and A_n denote the frequency and amplitude of the nth tone, and N is the number of the detected tones.

Stochastic synthesis

As illustrated in the last section, the wavelet coefficients of the envelope of the residual part are calculated after the spectral subtraction based on WT. Thereafter, the inverse CWT is employed to reconstruct residual signals in the time domain. The envelope plays as a filter and is convolved with white Gaussian noise to synthesize the stochastic component [64].

2.5. Evaluation criteria

Locations and signals of the sources are two of the most important information which must be taken into consideration for the auralization as mentioned in the introduction. Beamforming is first applied to calculate the locations of potential sound sources, and then the beamforming outputs with focusing on the angle steering to the source are regarded as the reconstructed source signals. Therefore, the first criterion regarding localization is the deviation between the calculated and the real source positions. Second, for signal reconstruction, the criterion will be the deviation between the reconstructed and original signals.

Ideally, listening tests are able to provide straightforward human perception on sound. However, listening tests are subjective and rely on the performance of the subjects during the tests. Statistical performance can be achieved, but it requires more participants and much time-consuming. Objective evaluation has the advantage of quick evaluation and being objective due to the comparison using physical or psychoacoustic properties [14]. For a periodic signal, the reconstruction can be evaluated by the amplitudes and frequencies of the tones, whereas for other signals, the spectra (or envelope) need to be compared to assess the reconstruction.

2.5.1. Localization error

Define $[\hat{x}, \hat{y}]$ as the coordinates on the reconstruction plane. The reconstruction plane Ω can be defined as reference to calculate beamforming outputs. The plane is divided into grids for the array to scan potential sound sources (Fig. 4.1). Each grid point on the reconstruction plane represents a focus point, where the sound source is probably located.

The z coordinate is omitted since the source locations and the moving trajectory are only on Ω on the xy-plane in the following analysis. The localization error as e_{loc}:

$$e_x = |x(n) - \hat{x}(n)|, \tag{2.36}$$

$$e_y = |y(n) - \hat{y}(n)|, \tag{2.37}$$

$$e_{loc} = \sqrt{{e_x}^2 + {e_y}^2}, \tag{2.38}$$

where $[x, y]$ is the original coordinates of the source.

2.5.2. Signal reconstruction error

Periodic signal

Define $\hat{f}(n)$ and $\hat{L}(n)$ as the reconstructed frequency and level of the beamforming output, respectively. The corresponding errors e_x and e_y are calculated by

$$e_f = \frac{|f(n) - \hat{f}(n)|}{f(n)} \times 100, \tag{2.39}$$

$$e_L = |L(n) - \hat{L}(n)|, n = 1, 2, ..., N, \tag{2.40}$$

If the two signals activate no different processing and perception in our hearing system, the synthesis can be regarded realistic and can be applied for auralization. Therefore, just-noticeable variations in frequency and amplitude are introduced.

Just-noticeable difference in frequency (JNDF) is approximated by [66]

$$JNDF = 0.002 \times f(n), n = 1, 2, ..., N. \tag{2.41}$$

for $f(n) > 500Hz$. Therefore e_f is denoted in the form of percent error in Eq. 2.39.

Just-noticeable difference in level (JNDL) is level dependent and the value decreases as sound pressure level (SPL) increases. For example, JNDL decreases from 2 dB to 0.7 dB as the source level increases from 30 dB to 70 dB SPL [66]. However, since the measure of JNDL depends on the measurement technique, there is no determined calculation for it. Therefore, the feasibility of level reconstruction for auralization should be evaluated with respect to different applications.

Spectrum

Jagla et al. [20] introduced a spectral method to evaluate the similarity between signals with accounting for human hearing by considering A-weighting and logarithmic sensitivity. They derived the difference between two signals by calculating the square of the subtraction between the absolute pressure values and then obtaining the logarithm of the square. However, log spectral distance which describes the difference between the logarithms of the two signals instead, has been applied to measure the perceptual distortion of speech processing [67]. It indicates that the logarithmic difference could be a criterion to quickly evaluate the perceptual difference. Therefore, the absolute logarithmic error is directly evaluated in the decibel scale as

$$e_S(f(k)) = |20log_{10}\frac{|\hat{S}(f(k))|}{|S(f(k))|}|, k = 1, 2, ..., K \tag{2.42}$$

where S and $\hat{S}$ are the Fourier transform of the original and reconstructed signals, K is the number of the frequency bins. With adding A-weighting to account for human hearing as in[20] and taking the average over the whole frequency range, e_{rec} is derived for the error estimation of the reconstructed signal:

$$e_{rec} = \frac{1}{K}\sum_{k=1}^{K}(e_S(f(k)) + w_A(k)). \tag{2.43}$$

e_{rec} is the measure used as the measure in this study to assess the perceptual difference between the reconstructed and original signals, or simply put, the reconstruction error. The small e_{rec} is, the more similar the reconstructed signal as the original signal, and the larger the reconstruction error is. Listening tests would be preferable to measure human perception more straightforwardly. However, the aforementioned measure is advantageous to quickly estimate errors in terms of various parameters. Listening tests could be included for future work to obtain more precise comparison from humans, and the results could also provide a correlation between subjective and objective measures.

2.6. Summary

In this chapter, the fundamentals of moving sound sources, delay and sum beamforming (DSB) and compressive beamforming (CB), spectral analysis and synthesis and evaluation criteria for localization and signal reconstruction have been introduced. DSB and CB are modified for moving sound sources. For DSB, the recorded signals are de-Dopplerized to be provided as the inputs. For CB, TDTFs form the sensing matrix to solve the ℓ_1-norm problem. Spectral analysis and synthesis are subsequently introduced to further parameterize the reconstructed time-domain signals in the frequency domain. Last but not the least, the evaluation criteria are proposed for the assessment of localization and reconstruction of tonal and broadband signals.

A block diagram of the methodologies of this thesis is shown in Fig. 2.8.

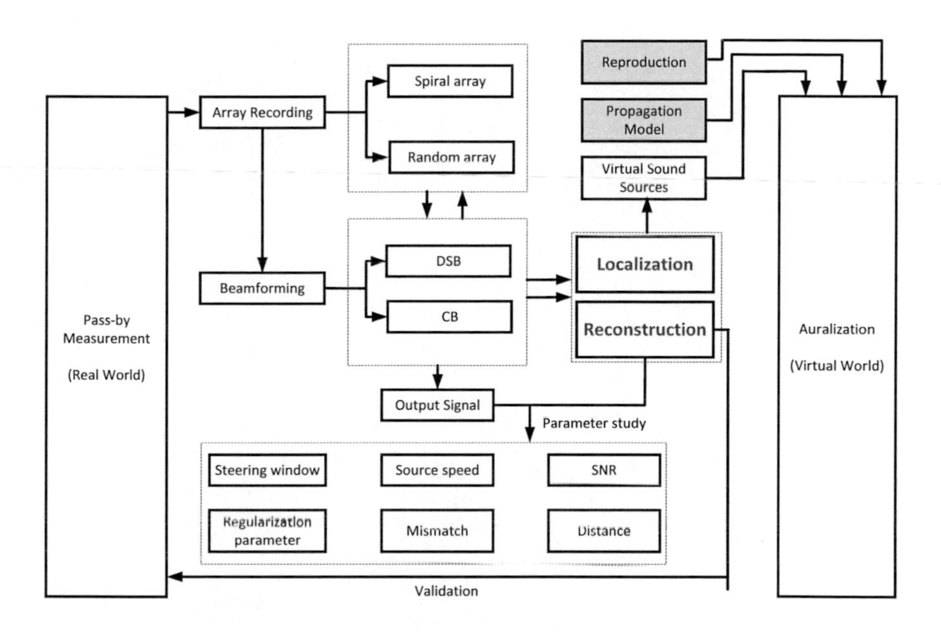

Figure 2.8.: The methodology block diagram for the localization and signal reconstruction of moving sound sources. The gray blocks are not conducted in this thesis, but they are the other two key components to be combined with the proposed source modeling to auralize moving sound sources.

3

Microphone arrays

To perform beamforming algorithms, microphone arrays are critical since parameters such as resolution and frequency range are determined by the array configuration. For compressive beamforming (CB), particular requirements should be fulfilled through controlling the array configuration. This chapter introduces several types of microphone arrays and their characteristics, and the design of arrays which will be applied in the following simulations and measurements.

3.1. Regular arrays

The regular array consists of regularly-spaced microphones, e.g. uniformly linear, cross and grid arrays [68, 69]. An example of a uniform linear array can be found in Fig. 3.1, which shows a vertical linear array with uniform deployment of microphones in sound fields with plane and spherical sound waves. Due to the periodic structure, grating lobes with the same magnitude as the main lobes occur. The side lobes deteriorate the localization ability of the array [27].

A preliminary pass-by measurement with a uniform linear 24-microphone array was applied on two pass-by trains in Appendix B [70]. The resolution of the applied linear array is limited at low frequencies. The Doppler effect was included in the recordings which were directly applied as input signals in delay and sum beamforming (DSB). De-Dopplerization could not be used due to the vertical structure of the array, resulting in the loss of ability to steer horizontally. Horizontal linear arrays can be used if the vertical resolution is of no interest. Therefore, in order to eliminate the Doppler effect and increase the spatial resolution, two-dimensional (2D) arrays are necessary for the localization and signal reconstruction of moving sound sources.

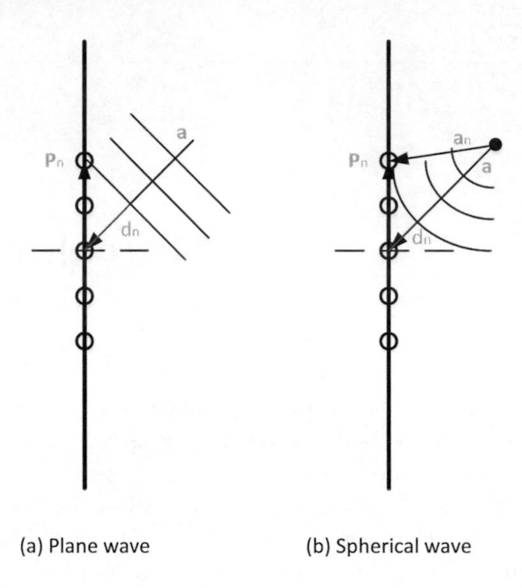

(a) Plane wave (b) Spherical wave

Figure 3.1.: Linear microphone array recording for the plane and spherical waves.

Fig. 3.2 illustrates two 2D regular configurations: grid and cross arrays. Like the regular linear array, ghost images caused by grating lobes at frequencies above the maximum frequency of the array are unavoidable. The occurrence of grating lobes due to the periodic distances between microphones is the main limitation of the regular array [38].

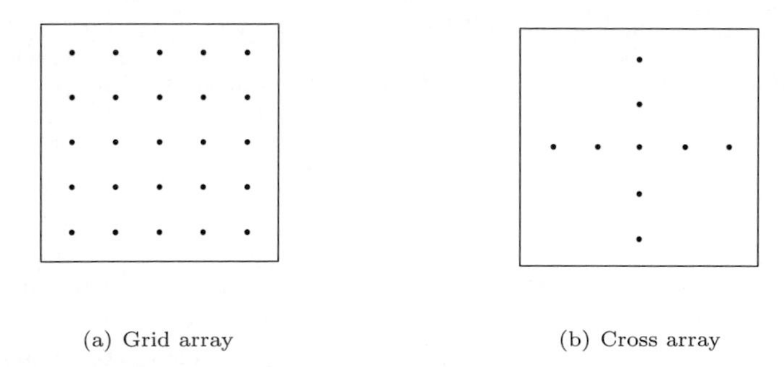

(a) Grid array (b) Cross array

Figure 3.2.: Examples of 2D regular microphone arrays. (a). Grid array; (b). Cross array.

3.2. Irregular arrays

Irregular arrays outperform the regular arrays because they overcome the limitation due to grating lobes [71]. It is necessary to apply appropriate configuration design schemes to avoid designing by numerous trials.

3.2.1. Spiral arrays

The spiral array, containing logarithmic and Archimedean spirals, is a simple irregular configuration which can be designed by only several parameters [38]. The spiral array shows a high concentration of microphones in the center and a gradual increase in microphone distance towards the end of the array aperture. The larger diameter guarantees better spatial resolution, and a higher maximum frequency can be exploited because of the small minimum microphone distance. Compared to regular arrays, spiral arrays have the advantages of decreasing MSL and avoiding grating lobes [38]. The microphone position $[x, y]$ on the logarithmic and Archimedean spiral apertures can be calculated as

$$[x, y] = [ae^{b\theta}cos(\theta), ae^{b\theta}sin(\theta)], \tag{3.1}$$

and

$$[x, y] = [(a + b\theta)cos(\theta), (a + b\theta)sin(\theta)], \tag{3.2}$$

where a, b are constant values and θ is the angle of circular rotation.

3.2.2. Sparse arrays

If the linear configuration is still taken into consideration, sparse arrays can be created by several criteria, e.g. nunredundant array by Vertatschitsch et. al. [72], coprime array [73] and sparse array [74]. To extend the arrays to 2D, the sparse array [74] and separable array [75, 76, 77] based on the nunredundant configuration have also been studied.

Spiral arrays of identical resolution provide lower side lobe levels than separable arrays. The merit of the above mentioned grid-like sparse arrays is the ability to employ the kronecker array transform (KAT) [78, 75] to accelerate deconvolution, which is able to reduce the sidelobe levels considerably [79, 31]. A separable array is designed and compared with the spiral array in section 3.2.1 in Appendix C [80].

3.2.3. Random arrays

Apart from the irregular arrays mentioned above, random arrays are another commonly used type of irregular arrays. They are especially suitable for CB due to their fulfillment of the incoherence requirement [53]. A scheme of designing pseudorandom arrays is developed in the next section.

A random array is known as an array with randomized positions of microphones. In this way, periodicity is avoided and grating lobes are constrained [81]. However, the difficulty of how to determine the randomized microphone positions on the aperture still remains [68]. Therefore, it is necessary to optimize the array configuration according to different requirements.

3.3. Array design

In this thesis, spiral and random arrays are utilized for DSB and CB, respectively. The spiral array can achieve good resolution for DSB, and the random configuration is a prerequisite for assuring an incoherent sensing matrix for the sake of CB. Due to the limitation of resources and practical constrains, the number of microphones in this thesis is limited to 32.

3.3.1. Spiral array design

In this thesis, an Archimedean spiral array with 32 microphones is designed. The array has six arms, four of which are equipped with five microphones while the other two hold six microphones. The array is 0.43 m high and 0.5 m wide. The minimum frequency with a steering angle of 30° is around 800 Hz according to $f_{min} \approx c/D$ [38], where c is the speed of sound and D [m] is the size of the array aperture. Fig. 3.3 illustrates the configuration of the array [77].

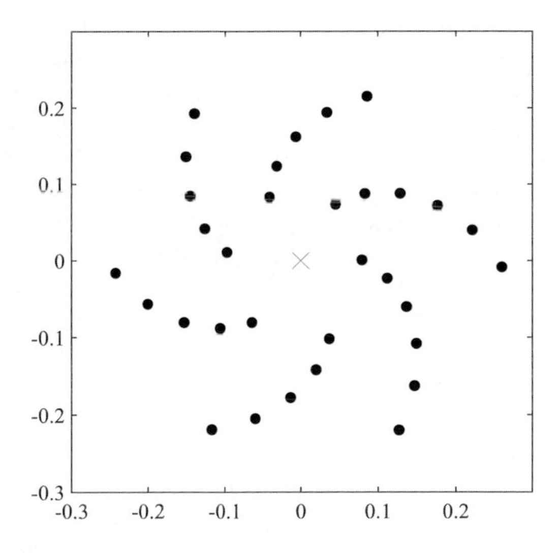

Figure 3.3.: The designed spiral microphone array. The "●" represents the position of a microphone and the "×" represents the origin of the array. The unit of the axes is meter.

The basic parameters and the configuration are given in Tab. 3.1.

N	D [m])	d [m]	L [m]	Resolution [m]	
				f [kHz]	Steering angle [°]
				3	30
32	0.5	0.04-0.06	1.5	0.64	

Table 3.1.: Several parameters of the spiral array setup. N is the number of microphones, D is the array diameter (length), d is the microphone spacing and L is the distance between the array and the near-side surface of the train.

3.3.2. Pseudorandom array design

As mentioned in the introduction chapter, the columns of the sensing matrix $\mathbf{H}$ should be incoherent to utilize CB. A random array is able to guarantee incoherence in the sensing matrix [53], and the RIP should be satisfied [54]. Gaumond et al. [82] proposed statistical restricted isometry property (StRIP) to help design sparse arrays. However, since DSB is also used as a comparison to CB, its performance should also be taken into account. Allowing for DSB with the same array also reduces the possible array configurations with meeting the requirements of CB. Gerstoft et al. introduced convex optimization to enhance the performance of beam patterns of 2D random arrays [83]. Good resolution and minimal maximum sidelobe level (MSL) were also used as criteria to design planar random arrays [81, 84, 85]. A framework for the design and optimization of 2D pseudorandom microphone arrays which benefits both CB and DSB by considering RIP and beam patterns is proposed next.

Design concept

If the positions of microphones on an array aperture are randomized, it would probably lead to the microphones clumping in a small area, and thus reducing the spatial resolution [81]. This would also increase the coherence of the sensing matrix because of very similar $R(t)$ of the closely localized microphones. Therefore, it is necessary to introduce restrictions to the randomization in the design of random microphone arrays, which leads to pseudorandom microphone arrays.

According to Kook et al. [81], segmenting an array aperture into units can guarantee that the microphones are well distributed on the array aperture to avoid clumping. Therefore, a baseline filter method is introduced to ensure the scattered distribution of the microphones [84]. Here, the baseline is defined as the distance between two arbitrary microphones in the microphone array [81]. The unit partition is able to deliver good localization performance for DSB. In this sense, the resolution and MSL are chosen as two of the design criteria.

Recalling Eq. 2.18, RIP in the current work is written as

$$(1-\delta_p)\left\|\mathbf{S}\right\|_2^2 \leq \left\|\mathbf{HS}\right\|_2^2 \leq (1+\delta_p)\left\|\mathbf{S}\right\|_2^2. \tag{3.3}$$

The resolution, MSL and RIP are the criteria to optimize the design of pseudorandom microphone arrays. Fig. 3.4 exhibits the proposed framework. The details are elaborated in the following contents.

Unit partition

The array aperture is discretized into 32 units, each containing a single microphone. According to Zheng et al. [84], each unit has eight possible positions, as illustrated in Fig. 3.5. In this figure, d represents the minimum distance between two microphones, and each microphone is $d/2$ spaced with the edges of the unit.

For a regular partition, it is possible that four microphones in four adjacent units locating on the corners are positioned next to each other, which is shown in Fig. 3.6(a). This distribution is not as scattered as the irregular unit partition in Fig. 3.6(b). Thus, the irregular unit partition is adopted for the further procedures.

Baseline filter

To design an array with M microphones, the number of all possible arrays generated following the idea of unit partition is 8^M. This is still a huge number, and thus further constrains need to be considered to reduce the number of trials.

The maximum number of possible baselines is $M \times (M-1)/2$ [84]. The definition of the baseline vector is defined as: a vector points from one microphone to another. Hence, the maximum number of baseline vectors is $M \times (M-1)$. So all the possible baseline vectors can be analyzed within a coordinate system that

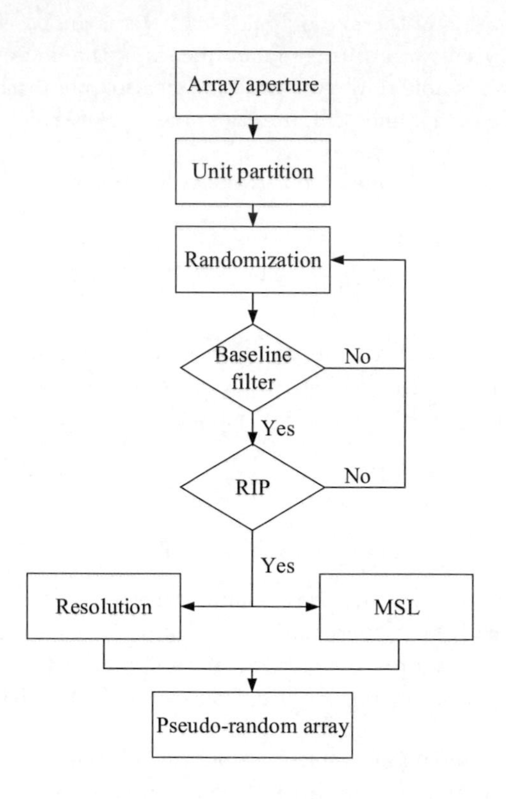

Figure 3.4.: A flow chart of the framework for designing and optimizing pseudorandom microphone arrays.

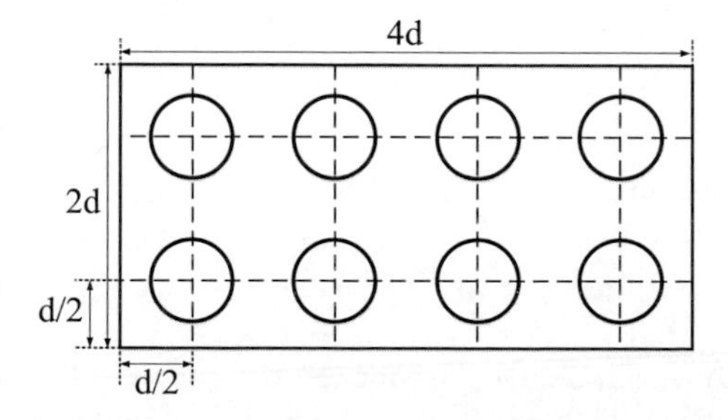

Figure 3.5.: A unit with eight possible microphone positions ("○").

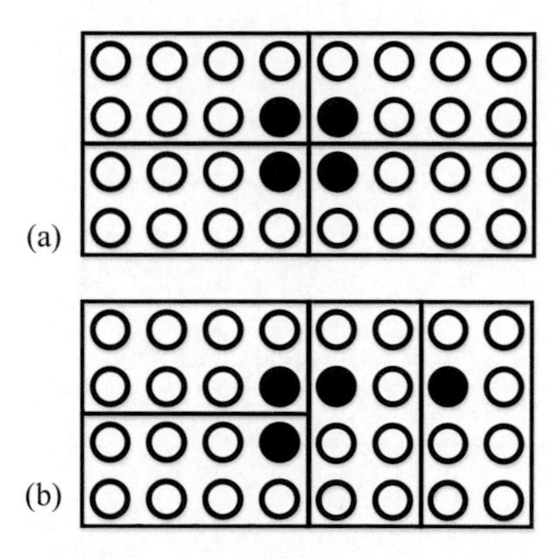

Figure 3.6.: Exemplary schemes of regular and irregular partitions based on four units. (a) is the regular unit partition and (b) is the irregular unit partition. The "○" represents the potential position of a microphone, and the "●" represents the real position of a microphone.

is created by possible baseline vectors and it is shown in Fig. 3.7 [84]. The initial points of the vectors are on the coordinate origin.

Define a square whose side length equals to $b = 2d$. If the array dimension is $H \times L$, the total possible number of squares $N_{sq} = max(H/b, L/b) - 1$. Similarly and sequentially, the side length of the ith ($i = 1, 2, ..., N_{sq}$) square $S(i)$ is $i \times b$, and $N_b(i)$ and $N_{bv}(i)$ represent the number of baselines and baseline vectors, respectively.

$K_{bv}(i)$ is the number of baseline vectors which have no intersection points with each other, and Fig. 3.8 gives an example of how to assign baseline vectors from array aperture to a square K_{bv}. Here, $N_{bv}(1) = 12$ and $K_{bv} = 4$ because there are only four baseline vectors (on the two opposite sides of the square) which do not have intersection points.

According to [84], the baseline filter is described by following equations:

$$\begin{aligned} N_{bv}(1) &\leq 4, \quad when\, M \leq 20, \\ K_{bv}(1) &\leq 4, \quad when\, M > 20, \end{aligned} \tag{3.4}$$

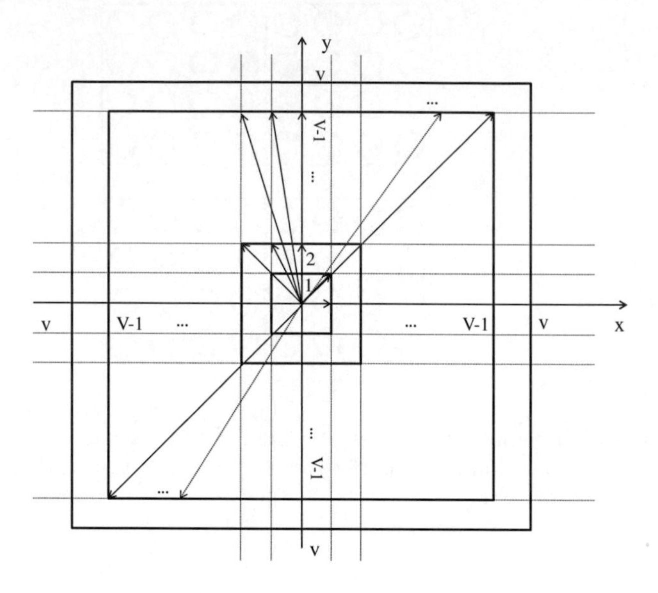

Figure 3.7.: Baseline vectors and squares.

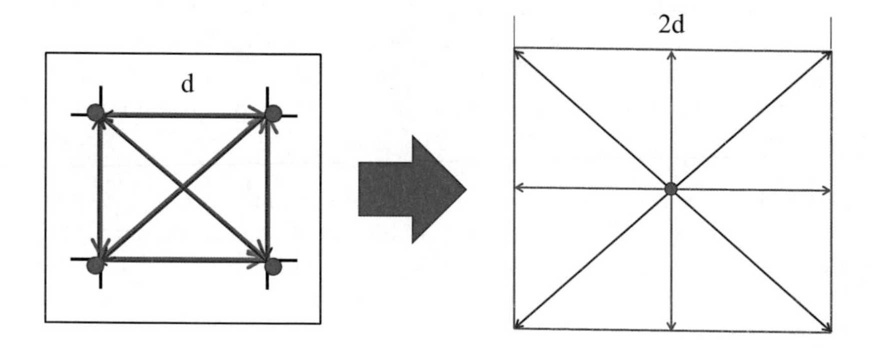

Figure 3.8.: An example of the calculation of K_1. In the left figure, the blue arrows are the vectors on the sides, and the red arrows are the vectors on the diagonals. The four circles on the corners are microphones; the baseline vectors are allocated in the first square as indicated in the right figure.

$$K_{bv}(2) \geq \begin{cases} 14, \ M \geq 14, \\ M, \ M < 14, \end{cases} \tag{3.5}$$

$$K_{bv}(N_{sq}) \geq 4. \tag{3.6}$$

Eq. 3.4 limits that there are at most two pairs of microphones located close to each other in the first (smallest) square. Eq. 3.5 assures that the microphones are distributed as scattered and uniformly as possible in order to improve the ability to suppress the sidelobe levels for DSB, and increase the incoherence of the sensing matrix for CB. Eq. 3.6 guarantees a bigger aperture which can deliver better resolution.

Pseudorandom array optimization

The arrays are designed on an aperture of 1.8 m × 1.8 m with 32 microphones. The array aperture could be larger to extend the frequency range to lower frequencies. However, large arrays are difficult to achieve due to practical reasons, e.g. construction and transportation difficulties. The locations of all microphones in every unit are randomized. Only if the configuration meets the requirement of the baseline filter method [84], can it be saved as a a possible array configuration. Following this rule, 1000 array configurations were generated. RIP was then tested and 50 arrays meeting the requirement remain. The resolutions of the 30° steering angle and MSLs of the 50 arrays are given in Fig. 3.9.

It can be observed that the resolution does not differ significantly along the simulations. Nevertheless, some of the MSLs deviate more than 10 dB. For DSB, high resolution and a low MSL are desired, but the two parameters are opposed with each other. Finally, the array with 19 dB MSL and 14° resolution with a 30° steering angle is selected as the optimized pseudorandom array. Fig. 3.10 shows the optimal array configuration among those being investigated. The aperture is partitioned into irregular units.

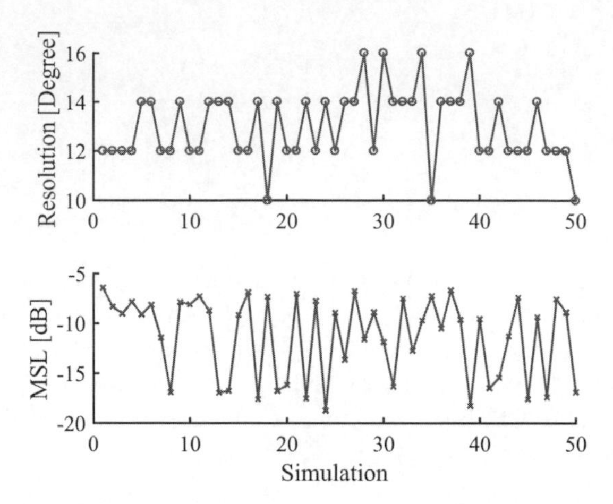

Figure 3.9.: The MSLs and resolutions of the 50 filtered arrays after the baseline filter and RIP test.

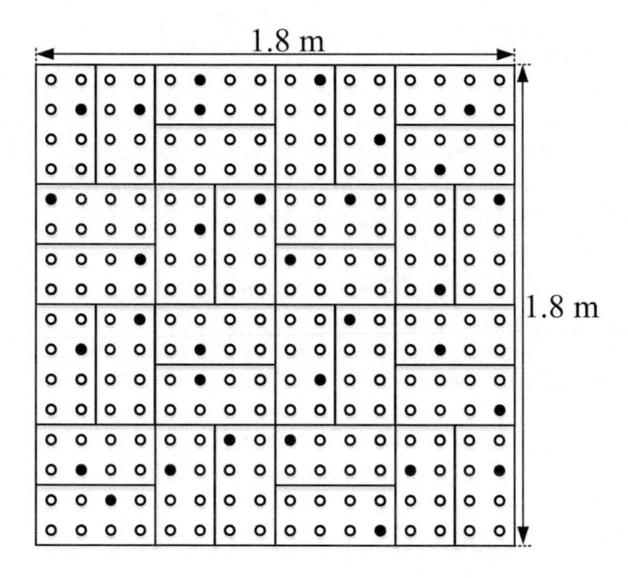

Figure 3.10.: The configuration of the optimized array with the scheme of irregular unit partition. The potential positions of the microphones are indicated by the "○", and the real positions of the microphones are represented by the "●".

3.4. Array construction

The designed arrays were constructed by the workshop at the Institute of Technical Acoustics (ITA), RWTH Aachen University.

3.4.1. Spiral array construction

The microphone positions were first marked on a hard board. 32 holes were drilled on the board such that the *Sennheiser* KE4 211-2 microphones could be inserted into the holes and held tightly. The threads twined around the capsules with proper force for clamping, and the ends of the threads were tied to the aluminum frame 3.11. The threads were glued and dried to be stiffer. After the thread nest totally dried, the hard board was removed. This way of constructing the array is quick, but it is only suitable for one configuration. Besides, the threads are easily tilted due to the weight of the cables.

Figure 3.11.: The spiral microphone array used in the measurement.

3.4.2. Pseudorandom array construction

The final design of the pseudorandom array is illustrated in Fig. 3.12. The array setup contains two layers with the array aperture on the front. 0.01 m × 0.01 m grids are formed by horizontal and vertical threads, thus allowing for plugging in microphones in arbitrary positions. On the back layer there are threads with coarser grids to hold the cables of the microphones. The two-layer setup is more stable for in-situ measurements.

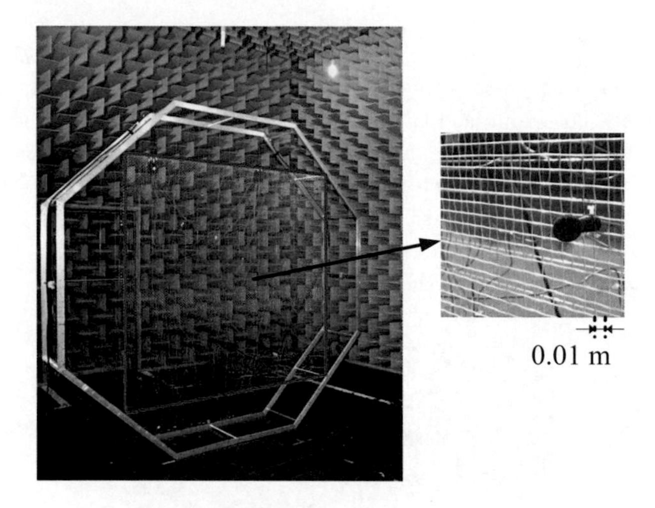

Figure 3.12.: The optimized pseudorandom microphone array. The area in the red square in the front layer is the array aperture. The spacing between the horizontal/vertical grids is 0.01 m.

The idea of meshing the array plane into 0.01 m grids with threads allows for arbitrary configurations within the size of 1.8 m × 1.8 m. However, the calculated positions of microphones have a high probability to not coincide with the positions of the real thread grids. It is thus necessary to recalculate the performance of the designed array with the real microphone positions. The nearest grid to plug in each microphone is found on the structure, and the comparison of the designed and real positions are given in Fig. 3.14. The deviations of MSL and resolution at 2 kHz are calculated. No change in resolution is found and the MSL only deviates by 0.1 dB. This implies that the beam pattern does not vary a lot due to the repositioning. According to the RIP test, the repositioned array still satisfies

Figure 3.13.: A side view of the optimized pseudorandom microphone array.

the requirement of the RIP. Therefore, the repositioning of the designed array slightly influences the usage of DSB but has no impact on CB.

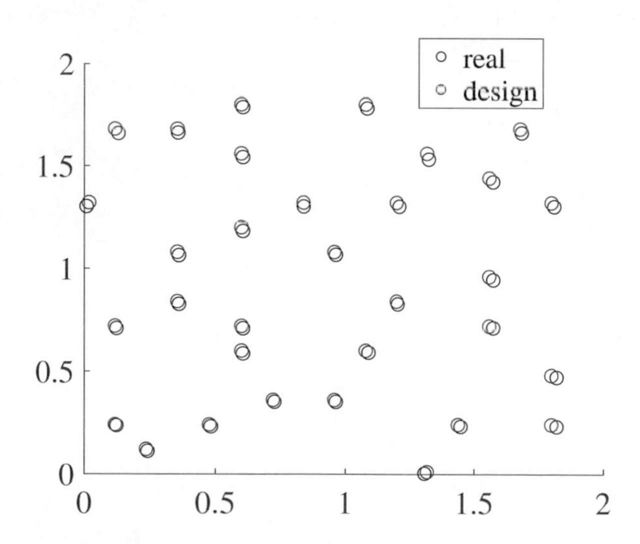

Figure 3.14.: The designed and real positions of the microphones of the pseudorandom array.

3.4.3. Microphones

The microphones used in this research are Sennheiser KE4-211-2. When the microphones are positioned in-situ, it should be confirmed that the front ends are on the same plane. During the outdoor measurement, wind shields are attached to the front end of the microphones to reduce the influence of the wind.

Although the structure of the pseudorandom array is stronger than the spiral array, the weight of the cables from the microphones still tilts the array aperture and thus causes uncertainties.

3.5. Summary

In this chapter, different microphone array configurations have been introduced. They can be classified in terms of dimension and regularity. A spiral array has been designed for DSB. An optimized pseudorandom array has been designed using unit partition, baseline filter, RIP test. The final selection accounts for good performance of DSB in terms of resolution and MSL. With the determined configurations, the spiral and pseudorandom arrays are constructed for the use in the following measurements.

4

Delay and sum beamforming

Chapter. 2 introduced the source model using the modified DSB for moving sources. The performance and capability of localization and signal reconstruction need to be investigated. Parameters such as the types of the input signals, source speed, parameters in beamforming (array geometry, time/spatial window, viewing window position etc.) can influence the performance of the model. This chapter will elaborate the evaluation of the model in several aspects by localizing and reconstructing a moving periodic sound source.

4.1. Simulation initialization

4.1.1. Steering window

When beamforming is applied to localize sound sources, the microphone array scans the plane by steering its angle, which is accomplished by introducing corresponding time delays to the microphones.

The viewing window [29], which is the product of the steering widow length and source speed, is defined as the gray area between the dashed lines in Fig. 4.1. Each grid point is processed within the viewing window. The time a grid point traveling within the window is t_{win}. To distinguish the spatial and temporal windows, the spatial window will be mentioned as viewing window and the corresponding temporal window t_{win} as steering window. The signal of each grid point will be utilized for source localization and signal reconstruction. t_{win} determines the spectral resolution. The spatial resolution depends on frequency [38], as well as the viewing window length [36]. Large viewing window enlarges the steering

*Most of the results presented in this chapter have been published in [6].

scope of the array and thus potentially worsens the spatial resolution, which has an impact on the localization accuracy.

Confining the viewing window between the two dashed lines, all grid points are processed after passing by this window. After all the grid points pass through the viewing window, the beamforming output signal is calculated at each grid point. For localization, the output signals are analyzed in 1/3 octave bands to study the localization error and detect sound sources on the moving plane. The source position is determined according to the results at the 1/3 octave bands of higher frequencies where localization is more accurate. With knowing the focus point representing the source position, the beamforming output calculated at this point is assumed as the source signal which will be discussed in the following subsection.

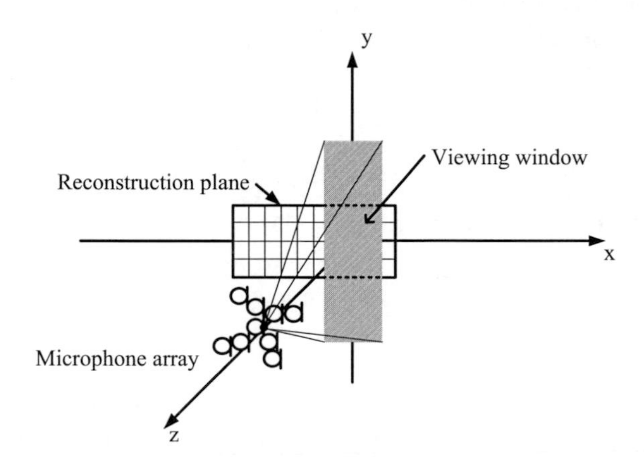

Figure 4.1.: The schematic of DSB applied on reconstructing moving sound sources. A microphone array is set away from the moving trajectory of a moving plane, which carries sound sources. The gray area between the two dashed lines represent the viewing window.

4.1.2. Source detection

In Chapter. 2 it was mentioned that the interpolated signal $\tilde{p}(t)$ is used instead of $p(t)$ during calculation. Recalling $t_e = t - \frac{R(t_e)}{c}$, in this thesis $t_e = t - \frac{R_e(t_e)}{c}$, it can be seen that the interpolation can only be proceeded with the knowledge of $R_e(t_e)$, which depends on the locations of the source and receiver. It contradicts to the purpose of obtaining the source location.

Therefore, the strategy to proceed the calculation is as follows. At t_1, the grid points on the vertical line L_1 as indicated by the solid dots in Fig. 4.2 form a group, and they are only processed when they pass by the viewing window, which is the spatial area between the two dashed lines [29]. The length of the viewing window is the product of the steering time window t_{win} and the source speed v. In each vertical line L_n, every grid point will be assumed as the source and thus there will be a set of interpolated received signals for all the 32 microphones. Subsequently, CB is applied on each point on L_n with the interpolated signals. The calculation continues point-wise on every point and vertical line, until all the points have been calculated. Fig. 4.2 exhibits the processing first at $[t_1, L_1]$, then at $[t_2, L_2]$. All the CB outputs are of the same length, t_{win}. Finally, a two-dimension matrix with the root mean square (RMS) values of the amplitudes of CB output signals are derived. Large RMS values are detected as potential sound sources, with the corresponding CB outputs as the reconstructed signals.

4.1.3. Zero-padding for harmonic signals

Most sounds generated by moving vehicles can be decomposed into deterministic and stochastic components. Spectral modeling synthesis (SMS) is an efficient approach to implement the decomposition [63]. The deterministic component can be represented by a sum of tones, and the stochastic component can be represented by filtered white noise. This way simplifies the sound representation and enables the parameterization and prediction of unknown sounds.

In this thesis, the main focus is on the deterministic component. A periodic signal is applied to test the performance of the model in a large frequency range. For a focus point, the recorded signals according to the emission time the point travels in the viewing window are cropped and rectangular window functions are applied. Zeros are added to the beamforming output signal before transforming the signal to the frequency domain using fast Fourier transform (FFT). The zero-padded signal is 1 second with the sampling rate of 44100 Hz. The magnitude of a particular tone after zero padding can be compensated by the ratio of the padded length and the original length (Equation 4.1).

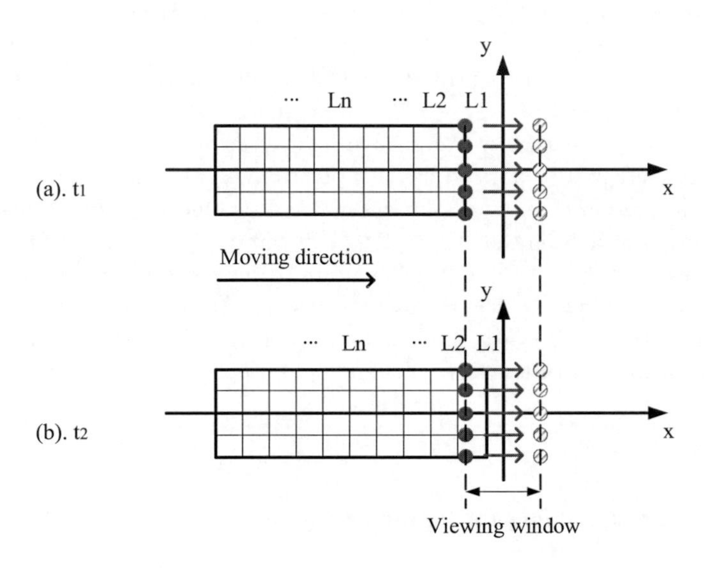

Figure 4.2.: The source detection procedure. The figure is the two-dimension view of Fig. 4.1 when looking in the $-z$ direction. The viewing window is the product of the time steering window t_{win} and the source speed v, and all grid points are only processed when they pass by the viewing window. The solid dots represent the grid points on a vertical line on the reconstruction plane. The arrows point to the positions of the grid points at the end of the spatial window. The grid points are piecewise processes from $[t_1, L_1]$ to $[t_n, Ln]$.

$$A_{ori} = \frac{N_{zp}}{N_{ori}} \times A_{zp}, \tag{4.1}$$

where A_{ori} and N_{ori} are the magnitude and the length of the original signal, and A_{zp} and N_{zp} are the magnitude and the length of the zero-padded signal. The level can be calculated as $L = 20log(A)$ with number 1 as the reference sound pressure.

4.1.4. Simulation setup

The simulation of pass-by measurements is implemented following Eq. 2.10. Reflections are neglected since it can be described by an image source and thus can be constricted by steering the array to the real source. Other propagation effects are also not included to simplify the model.

A moving plane with sound sources recorded by a stationary microphone array is simulated. The moving trajectory of the plane is in the x-direction with its center on the axis. The plane (5 m × 1.5 m) is meshed into 10 cm-grids. The spacing between grid points is smaller than the distance a grid point travels during t_{win} to avoid omitting any potential sound sources between grid points.

Three virtual sound sources, S1, S2 and S3 are placed on the plane and move simultaneously. S1 is in the middle of the plane, and S2 and S3 are located with the offsets of [-1 m, -0.55 m ,0 m] and [1 m, -0.55 m ,0 m] referring to S1. The initial position of S1 is $[-\frac{t_{run}v}{2}$ m, 0 m, 0 m], where t_{run} is the time duration of the plane's movement in the simulation. A periodic signal is regarded as the target source signal to be reconstructed. The frequencies in the signal range from 1 kHz to 8 kHz with 500 Hz step. This periodic signal is attached to S1. Meanwhile, a white Gaussian noise $n(t)$ is added to S1 as well. The spectrum of the periodic signal with $n(t)$ added is shown in Fig. 4.3. Interference noise sources with the same sound power as $n(t)$ are attached to S2 and S3, respectively. The sampling rate is 44.1 kHz in the following simulations and experimental analysis.

The simulation of microphone recordings of moving sound sources is according to Eq. 2.10. The MATLAB codes can be found in Appendix A.

The array to be used here is the spiral array designed in Section 3.2.1. The array origin is at [0 m, 0 m, -1.5 m]. The sketch of the simulation setup is shown in Fig. 4.4.

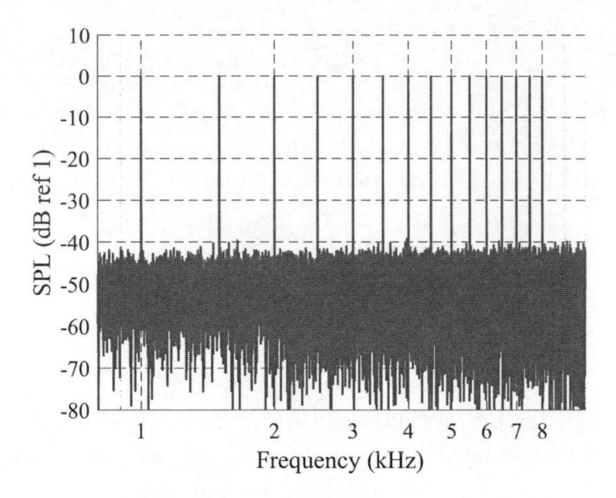

Figure 4.3.: The spectrum of the periodic signal attached to S1 with adding white Gaussian noise $n(t)$.

The entire simulation was repeated 1000 times to test the repeatability and no deviation was reported.

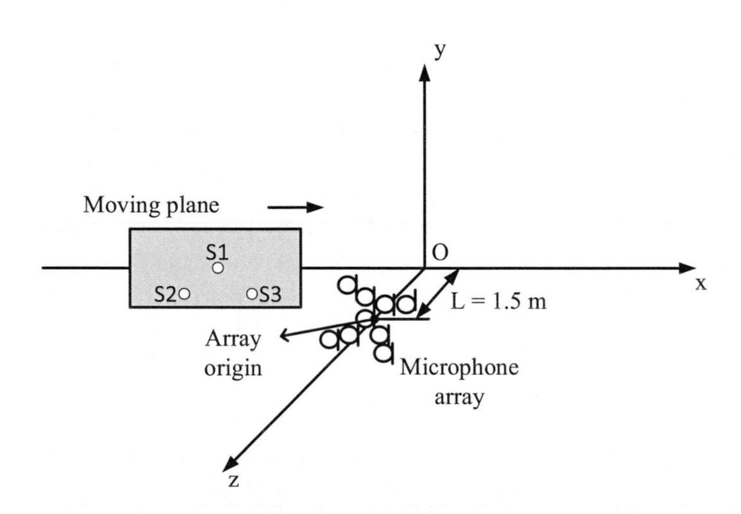

Figure 4.4.: Sketch of the simulation of a moving plane with three sound sources measured by the spiral microphone array.

4.2. Model evaluation

There are several parts in the calculation of the model that can lead to errors. First of all, spline interpolations are applied during the simulation of the pass-by sound source recording and de-Dopplerization. In addition, although zeros are added to the steered signal, the reconstructed signal still suffers from wrong frequency reconstruction, and hence also wrong amplitude reconstruction. It correlates to the length of the steering window.

Furthermore, the properties of the microphone array also play a significant role in the error estimation. The length of the viewing window in Fig. 4.1 is the product of the steering window and the source speed. The larger the viewing window becomes, the larger the steering angle of the array gets, thus enlarging the beam width and worsening the spatial resolution.

The following evaluation will be focused on quantitatively analyzing the errors in terms of several parameters. In terms of the localization error, the results are reported at six 1/3 octave bands, i.e. 1 kHz, 2 kHz, 2.5 kHz, 4 kHz, 5 kHz and 8 kHz.

4.2.1. Steering position

In [31], reconstruction at different steering positions was discussed. The results showed that steering the array to 0° can achieve better reconstruction of the signal, including localization, the reconstruction of frequency and amplitude, compared to -26° and 26°. For the 0° case, the origin of the coordinate system O overlaps with the center of the viewing window, as depicted in Fig. 4.1. In this chapter, the discussion of the viewing window is extended to another two cases, placing the origin O either to the left or right boundary of the viewing window, leading to two additional steering positions, which can be mentioned as left and right windows (Fig. 4.5).

Note that the “moving” microphone and the source are assumed stationary when the source is on the y-axis, as mentioned in Chapter. 2 that in this case $R^0 = min(R(t))$.

In the following simulations, the length of the steering window is 1024 samples, and the speed of the moving plane is 20 m/s. The simulation results are shown

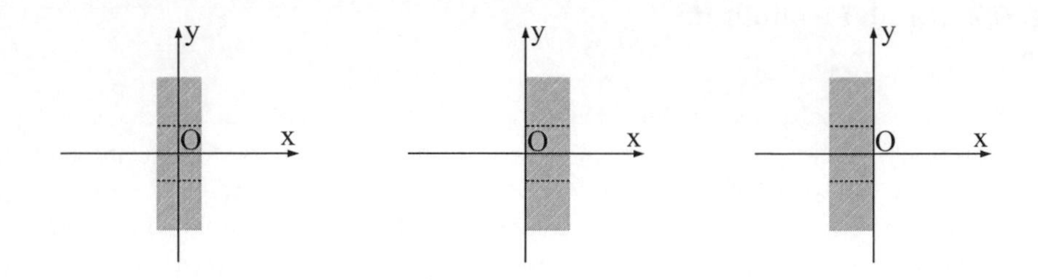

Figure 4.5.: Three positions of the viewing window: middle, left and right. The gray area represents the viewing window.

in Fig. 4.6. Regarding localization, no error in the y direction is detected. Hence Fig. 4.6(a) only shows the error in the x direction in 1/3 octave bands with center frequencies from 1 kHz to 8 kHz. As can be seen, localization errors appear at the 1/3 octave bands withe center frequencies of 1 kHz, 1.25 kHz, 1.5 kHz and 2.5 kHz. The large errors at 1.25 kHz are due to the low energy in this band since no tone of S1 is included. As the localization ability increases with increasing the frequency and three tones are included in the bands of 5 kHz, 6.3 kHz and 8 kHz, no error is detected in this frequency range. Therefore, the localization result of the band of 5 kHz is selected as the position of S1 in the following contents, and the beamforming output signal at this position is considered as the reconstructed signal.

For the frequency percent error in Fig. 4.6(b), all the errors are equal or smaller than 0.04%. Also, no obvious trend is observed with increasing the frequency. Nevertheless, the middle position delivers more stable results as it produces more zero errors and a lower upper limit around 0.03%. As for Fig. 4.6(c) with level errors, the error range is between 0 dB and around 2 dB. A slight but fluctuating increase as the frequency increases is illustrated. It appears that in Fig. 4.6(b) and Fig. 4.6(c), the left viewing window provides more fluctuating errors, including errors close to 2 dB at 7 kHz and 8 kHz in Fig. 4.6(c).

4.2.2. Steering window length vs. source speed

As the steering window length and source speed influence the frequency and spatial resolution, this section takes the two parameters into account for the error analysis. A middle steering position as shown in Fig. 4.1 is applied with varying

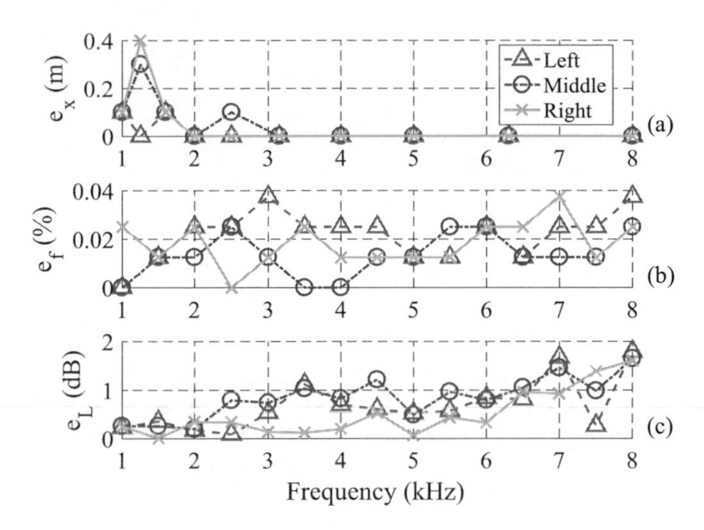

Figure 4.6.: The error of localization, frequency and level with varying the position of the viewing window. a). Localization errors in the x direction; b). Frequency percent errors; c). Level errors.

window lengths of 256, 512, 1024, 2048, 4096 samples, while varying the source speeds from 20 m/s to 120 m/s with a step of 20 m/s.

As no localization error is detected at the 1/3 octave bands with center frequencies higher than 4 kHz, Fig. 4.7 shows the mean localization errors (in x and y direction, respectively) of the five bands with one single tone, i.e. with center frequencies of 1 kHz, 1.6 kHz, 2 kHz, 2.5 kHz and 4 kHz. In the x direction, no obvious correlation between $\bar{e}_x$ and window length or speed is observed, except for the error increase of the 4096-sample window with increasing the speed from 80 m/s to 120 m/s. The errors can be explained by discussing several factors. Low frequency and large viewing window length as discussed in Section 4.1.1, errors introduced by de-Dopplerization [36] could decrease the localization accuracy. Additionally, spectral leakage exists in the calculation at every focus point, leading to energy at the frequency bin of the source leaking to other frequency bins. It could also result in localization error since the focus point with the highest energy is considered as the source. Therefore, the four factors together can explain the localization errors in Fig. 4.7. Only for large speed and large steering window length, the viewing window length becomes the dominant influencing factor since the localization error in the x direction is positively correlated with it.

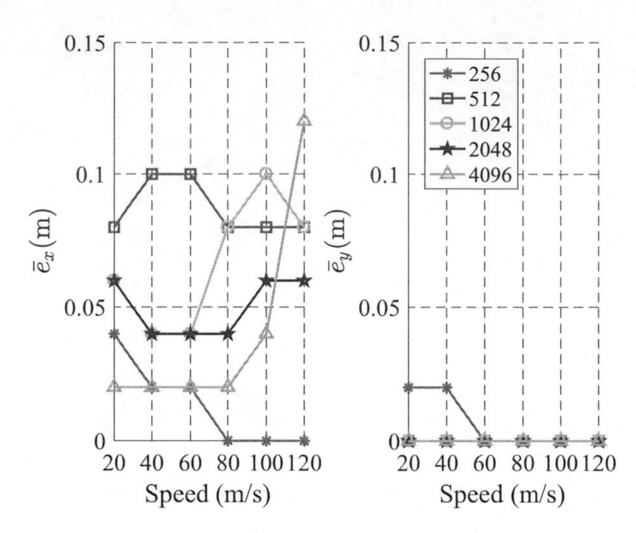

Figure 4.7.: The mean localization errors at x and y directions versus source speed for various windows with 256, 512, 1024, 2048, 4096 samples.

Fig. 4.8 shows the error bars of the reconstructed frequencies and levels versus source speed with various window lengths. In general, as can be seen, the maximum errors can reach 0.35% for frequency reconstruction and over 7 dB for level reconstruction. Furthermore, the mean and minimum errors of any single error bar are close to each other, which indicates that large errors are rare in the reconstructed tones in the periodic signal.

For the frequency reconstruction, the frequency error increases as the window length decreases and the source speed increases. Although zero padding adds more frequency bins, it is not able to increase the frequency resolution. A larger window length delivers better frequency resolution, thus allowing for better accuracy of frequency reconstruction. In addition, increasing the source speed introduces more errors to de-Dopplerization [36], thus leading to larger frequency errors.

For the level reconstruction, the mean error curves of the four small windows are almost flat over all the speeds, whereas the mean error of the window with 4096 samples remains quite similar for the speed from 20 m/s to 60 m/s while increase as speed increases from 80 m/s. Taking a look back at Fig. 4.7, it can be concluded that large localization error delivers large level error. The errors

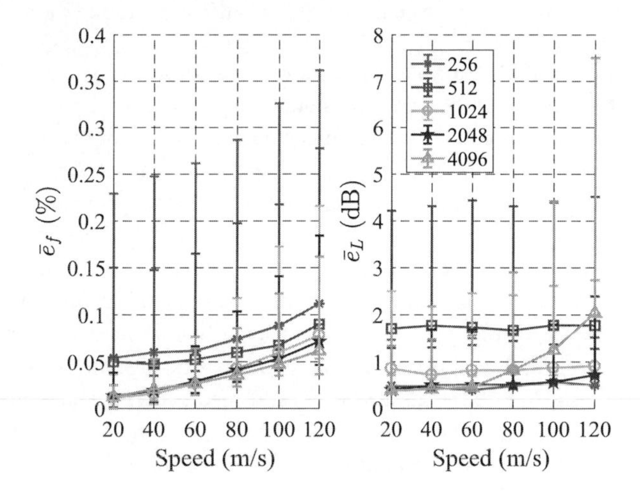

Figure 4.8.: The mean reconstructed frequency percent errors and level errors versus source speed for various window lengths. The marker in the middle of each error bar represents the mean error, the upper horizontal line represents the maximum error and the lower line represents the minimum error.

are influenced by the frequency resolution and spectral leakage, as well as errors from de-Dopplerization. It is uncertain how the factors vary along with the window length in terms of level error. Therefore, the level errors in Fig. 4.8 are not in correlated relationship with the window length and source speed. Only when the speed and the window length are comparatively large, e.g. 4096-sample window, the viewing window length becomes positively correlated with the localization error as it leads to the failure of assuring the spatial resolution and large errors from de-Dopplerization. In this sense, the varying tendency of the level reconstruction errors is similar as that of the localization errors in terms of speed and window length.

Most of road and track vehicles' speed is confined to lower than 100 m/s (about 360 km/h). In this speed range, Fig. 4.8 illustrates that the mean percent errors of the frequency reconstructions are below 0.1%, and level errors below 2 dB. Upper bounds above JNDF indicate that some of the frequency reconstructions can be perceived differently from the original signals. Similarly, although most of the mean level errors are below 1 dB, many upper bounds are still large, some of them above 4 dB. Therefore, it is not a trivial task to select the parameters in the model. Considerations must be taken in terms of different application scenarios.

Take the window size of 256 samples as an example and how the error changes as a function of frequency with five various source speeds can be found in Fig. 4.9. Peaks can be observed in the frequency error curves. The combination of poor frequency resolution, spectral leakage and errors caused by de-Dopplerization might contribute to the unexpected deviations. However, the overall frequency errors for all the frequencies are lower than 0.2%, except for those at 5 kHz. As for the level reconstruction, almost all the errors are below or at least around 1 dB. Therefore, the model is able to regenerate the given periodic signal with the reconstruction deviations of frequency and level below the JNDF and JNDL for most of the cases with respect to the frequencies of tones and source speeds. However, large biased signal reconstruction could also occur, for instance, in the cases where the upper bounds of the error bars appear in Fig. 4.8. It is thus necessary to conduct a detailed parametric study to optimize the parameters and reduce reconstruction errors.

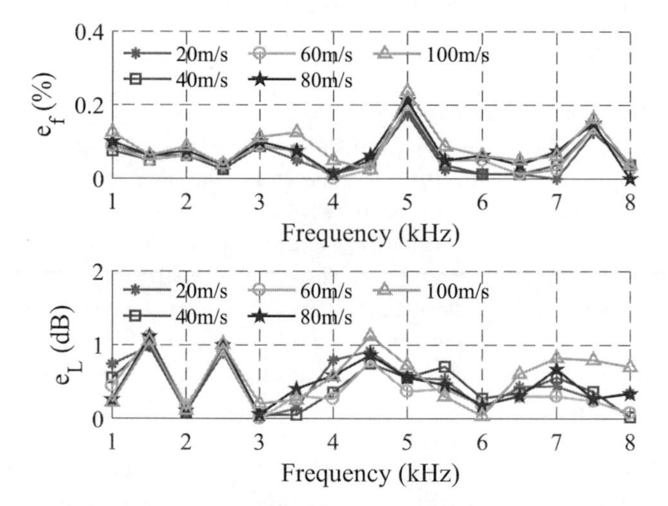

Figure 4.9.: The frequency percent errors and level errors versus source speed for various frequencies.

4.3. Application on a moving loudspeaker

4.3.1. Measurement setup

A pass-by measurement of a car using the spiral microphone array was conducted on the Proving Ground of Institute for Automotive Engineering , RWTH Aachen University. The proving ground has a 400 m long test track with an acoustical part which was built referring to ISO 10844/94 [86]. The array was 1.5 m away from the moving trajectory of the near-side surface of the car. A loudspeaker was installed and sealed on the back window frame. Referring the front-bottom point of the near-side plane of the car as [0 m,0 m], the loudspeaker was approximately located at [3.1 m, 1.2 m]. The loudspeaker played the periodic signal which was used in the simulation in order to validate the model. For all the pass-by measurements, the car was instructed to run on a straight line with constant speeds, i.e. 30 km/h, 50 km/h, 80 km/h and 100 km/h with two repetitions.

A set of photoelectric sensors were placed between the trajectory and the array. The sender was placed on the other side of the car's trajectory. The emitted infrared light could be subsequently received by the receiver sensor (can be seen in Fig. 4.10) with the absence of obstacles. A switch of receiving the light at the

Figure 4.10.: Setup of the pass-by measurement. The spiral microphone array and the photoelectric sensor were set 1.5 meter from the near-side surface of the car.

sensor would generate an impulse in the sensor's recording channel. Two impulses were excited due to the car approaching and leaving. The receiver sensor was connected to the same preamplifier with the microphones and thus synchronized. The pass-by time of the car is $[T_{Apr.}, T_{Lea.}]$. Note that $[T_{Apr.}, T_{Lea.}]$ is in terms of the emission time. Following what was shown in Fig. 4.2 in Section. 4.1.2 and taking the near-side surface of the car as the reconstruction plane Ω), the loudspeaker can be localized and its signal emitted during passing by the steering window can be reconstructed.

Besides, the car's speed can also be derived from the division of the car length and the pass-by time recorded by the sensor. Fig. 4.10 shows the pass-by measurement setup. The temperature θ during the measurement was around 21.9°C, thus the sound speed is 344.3 m/s.

4.3.2. Measurement results

The reconstruction plane is divided into grids with 5 cm $\times$ 5 cm resolution. The localization results at four selected 1/3 octave bands with center frequencies of 2 kHz, 2.5 kHz, 4 kHz and 5 kHz are shown in Fig. 4.11. The speed of the car was 50 km/h. It is clear that the resolution increases with increasing the frequency. Therefore, the localization of the loudspeaker also differs between frequencies. To simplify the validation, the loudspeaker is considered as a point source. The loudspeaker localization is sought in the six 1/3 octave band as mentioned before in each measurement with different speeds.

Since the energy of the beamforming output is the benchmark to identify if the array is steered to the real source, the location with the largest output energy is determined as the correct location for each measurement. As a result, the location of the loudspeaker is between [3.06 m, 1.17 m, 0 m] and [3.12 m, 1.20 m, 0 m]. With the determined locations, the signal of the loudspeaker is reconstructed and compared with the reference signal which was measured in the hemi-anechoic chamber. Two window lengths are selected: 256 and 1024 samples. The corresponding frequency percent errors and level errors are illustrated in Fig. 4.12 and Fig. 4.13.

With the determined locations, the signal of the loudspeaker is reconstructed and compared with the reference signal which was measured in the hemi-anechoic chamber. Two window lengths are selected: 256 and 1024 samples. The corresponding frequency percent errors and level errors are illustrated in Fig. 4.12 and Fig. 4.13.

For the 256-sample window, the frequency percent errors at 1 kHz and 2 kHz are much larger than those at other frequencies. The limited sample number

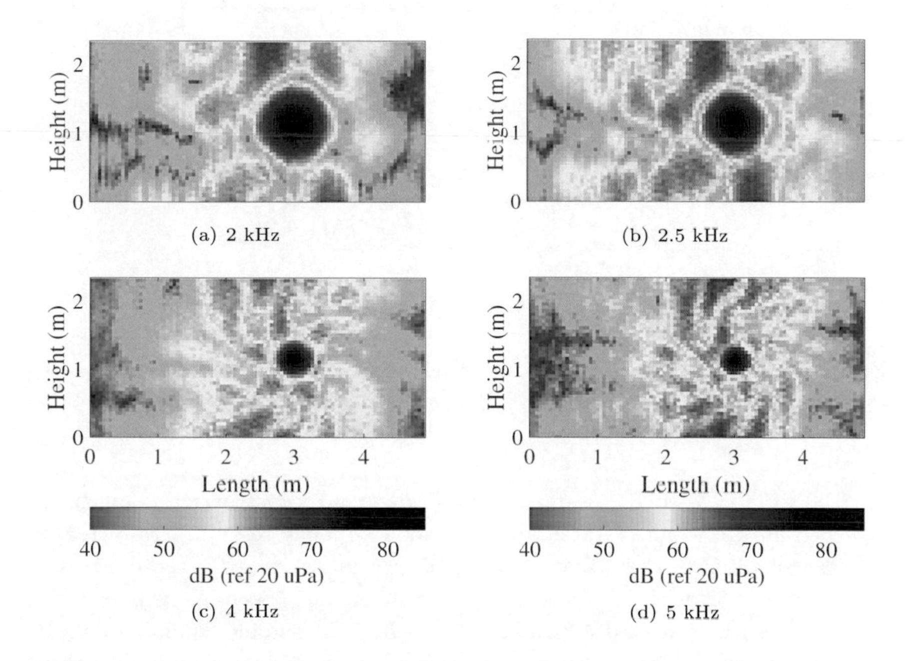

Figure 4.11.: Localization results on the near-side surface of the car at the speed of 50 km/h. The center frequency of the 1/3 octave band in each figure is: (a). 2 kHz; (b). 2.5 kHz; (c). 4 kHz; (d). 5 kHz.

lead to wrong localization especially at lower frequencies. The errors of level reconstruction are all below 4 dB.

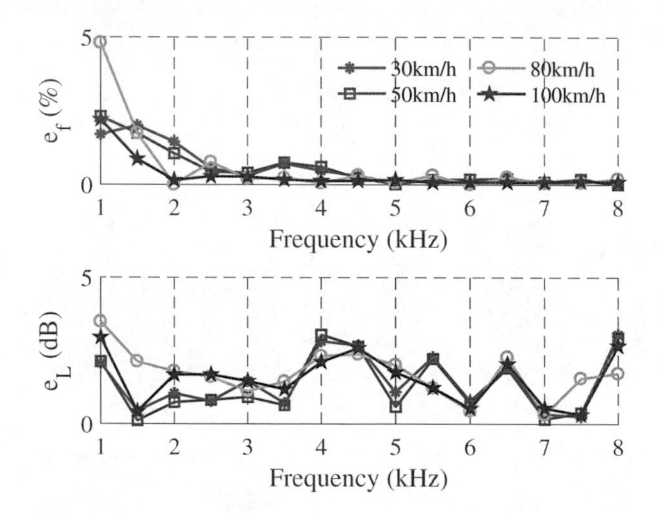

Figure 4.12.: The frequency percent errors and level errors of the reconstructed tones of the periodic signal versus source speed for various frequencies with window length of 256 samples.

For the window with 1024 samples, the frequency reconstruction is much less biased than the window stated above as the frequency resolution is increased. Additionally, it is apparent that the percent error increases as the speed increases. The level reconstruction is comparable to that of the 256-sample window. The maximum error is around 4 dB. Therefore, the 1024-sample window would be preferable to choose for auralization to achieve a more similar hearing perception as the original reference signal.

Compared to simulation, the results of the measurements have more deviations according to some uncertainties. First of all, biased microphone positioning existed in the measurements. The car's trajectory could also not exactly follow the desired track. Those factors lead to wrong calculation of delays and thus increase errors. Furthermore, the uncertain acoustic center of the loudspeaker, reflections, wind etc. could all bias the results.

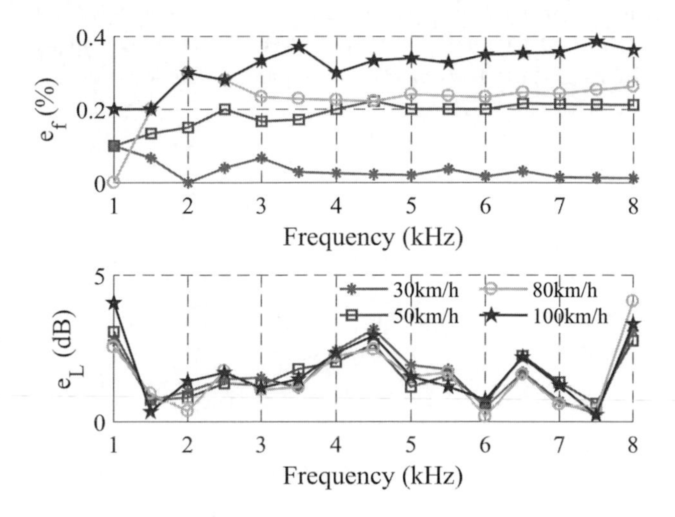

Figure 4.13.: The frequency percent errors and level errors of the reconstructed tones of the periodic signal versus source speed for various frequencies with window length of 1024 samples.

4.4. Summary

This chapter explores the reconstruction of a moving periodic signal based on DSB using the spiral array. The simulation results are evaluated by studying the reconstruction errors of localization, frequency and level by varying the parameters included in the model, i.e. steering position, steering window length and source speed. The steering window and the source speed have significant influence on the localization and reconstruction accuracy of the model. The middle position of the steering window performs better than the left and right positions according to the accuracy analysis of localization and signal reconstruction. In addition, the localization accuracy could be degraded by large windows and low frequencies, spectral leakage and errors caused by de-Dopplerization. For signal reconstruction, the frequency reconstruction has negative correlation with the window length and source speed. The error of level reconstruction shows similar varying tendency as the localization error in the x direction. It does not vary along with the window length and source speed, except that in the case of the largest window with 4096 samples, the errors appear to positively correlate with the two parameters at higher speeds. Various window lengths are necessary to be applied in terms of various source speeds to deliver more accurate results. The combination of limited frequency and spatial resolution, spectral leakage and errors caused by de-Dopplerization would result in unexpected deviations, such as the frequency errors at 5 kHz in Fig. 4.9.

5

Compressive beamforming

The source model has been tested by DSB by conducting error analysis and model evaluation on a moving periodic sound source. This chapter extends to model moving sources by applying CB by using the pseudorandom microphone array. An engine signal is attached to the moving sound source to test the broadband performance of the model. Several parameters are selected to evaluate the model through error analysis. The results of DSB are also given to compare with CB. Finally, a pass-by loudspeaker with the same engine signal was measured to validate the model.

5.1. Simulation setup

Two moving sound sources (S1 and S2) are simulated by Eq. 2.14 and Eq. 2.18. Note that the equations are denoted in continuous time, but discrete time is required in digital processing. Thus the calculated signals on the right-hand side of Eq. 2.14 are interpolated and resampled in terms of uniformly spaced time stamps to obtain $p(t)$ to simulate real recordings. S1 and S2 fixed in a plane which moves in the $-z$ direction at 20 m/s. The pseudorandom microphone array is placed 5 m away from the moving trajectory that is parallel with the array aperture. The moving plane is regarded as the reconstruction plane Ω, on which the beamforming calculations are conducted. Ω is meshed into grids and the distance between two grid points is 0.1 m. Each grid point is scanned as a potential sound source's position. S1 is on the origin of Ω, and S2 is 0.5 m located above S1 on the same vertical line. S1 and the origin of the array are both on the xz plane in the coordinate system. A sketch of the simulation can be found in Fig. 5.2. A 6 s recording of engine noise and a 6 s periodic signal are attached to S1 and S2, respectively. The periodic signal consists of a fundamental tone of 500 Hz and all its harmonics up to 8 kHz (with a random deviation on each harmonic of up to ±50 Hz), plus a 200 Hz tone. The spectra of S1 and S2 are shown in Fig. 5.1.

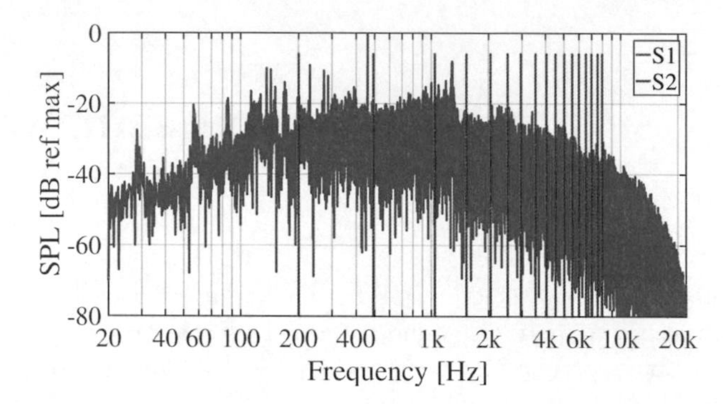

Figure 5.1.: The spectra of the two sound source signals S1 and S2.

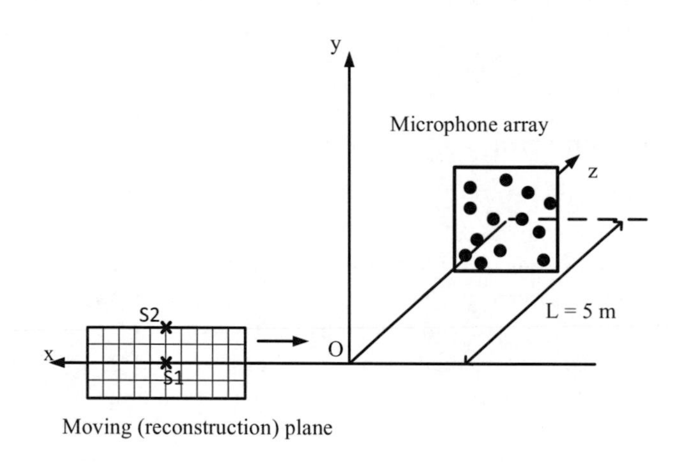

Figure 5.2.: The sketch of the simulation.

The duration of S1 and S2, and the moving time of Ω is 6 s. The starting and stop positions of Ω are symmetric in terms of the origin O of the x axis. S1 is the target source to be localized and reconstructed and S2 is regarded as an interference source. A wide frequency range of signal reconstruction can be studied by virtue of the broadband engine noise.

The simulation of microphone recordings of moving sound sources is according to Eq. 2.10. The MATLAB codes can be found in Appendix A. The source detection follows the procedures proposed in Section. 4.1.2.

5.2. Model evaluation

The proposed method using CB is evaluated by means of errors regarding localization and signal reconstruction. Regularization parameter, window length, SNR and mismatch are selected for the error analysis. Errors using DSB are also given to compare with CB.

5.2.1. Regularization parameter vs. window length

It is critical to select the regularization parameter as it determines the tradeoff between the fit of the solution to the original data versus the sparsity prior [44]. The selection of the regularization parameter still remains a difficult question, and trials through simulations were conducted to find out the optimal solution [44, 48]. It was suggested that a low noise level could be employed for the selection of regularization parameter to guarantee capturing all nonzero elements [47]. It was also pointed out that the regularization parameters in the constrained and unconstrained forms are related [44]. Therefore, the regularization parameter β in the Dantzig Selector [87, 45] is used as the search basis. $\beta = \epsilon_N \sigma$, where $\epsilon = \sqrt{2 \log N}$ (N is the number of the microphones) and σ is the standard deviation of the noise. Simulations are conducted in the neighborhood of β to search for a good choice of the regularization parameter λ in the unconstrained form in Eq. (2.30).

The errors of varying λ from 0.5β to 2β are studied. Additionally, the length of a steering window determines the spatial and spectral resolution, which has been discussed for DSB in Chapter. 4. Thus the regularization parameter and the window length are jointly investigated. The errors of localization and signal reconstruction are compared in Fig. 5.3(a) with SNR = 30 dB. Similar performance between DSB and CB can be observed except for some large variations for the windows with 32, 64 and 256 samples. For CB, most of the errors in terms of various λ achieve similar results. For the 64- and 256-sample window, no error is detected from the localization and signal reconstruction.

Now the SNR is decreased to 5 dB, a value more prone to be found in real measurement situations. The errors are shown in Fig. 5.3(b). e_{loc} and e_{rec} become larger with decreasing the SNR as expected, and the localization results of DSB and CB are still quite similar. Nevertheless, DSB and CB can be clearly distinguished in terms of signal reconstruction. For all the λ selected, the e_{rec} of CB are all below those of DSB, and each λ delivers a separate curve. The level difference could be quite large, e.g. around 6 dB for the case of 64-sample window with $\lambda = 0.5\beta$ and $\lambda = 1.75\beta$. The range of SNR is then extended to [15 dB, -5 dB] to have a better understanding of the influence of SNR (Fig. 5.4, $\lambda = 0.75\beta$). It can be seen that the reconstruction error increases as SNR decreases, and gradually CB outperforms DSB.

In Eq. 2.23, the reconstructed signal $\hat{s}(t)$ from DSB also contains noise, the incrementation of which would lead to increasing error in $\hat{s}(t)$. On the contrary, CB takes noise into account during the calculation as shown from Eq. 2.27 to Eq. 2.29. Thus CB outperforms DSB for signal reconstruction with the presence of strong noise, i.e. SNR = 5 dB in this study. It can be expected that both algorithms would perform similarly with the presence of slight noise (SNR = 30 dB), as in Fig. 5.3(a). However, the localization ability of CB shows no clear advantage over DSB in the current situation. The distance between Ω and the microphones is large compared to the microphone distances, which would result in potential coherence in the TDTF and cause errors. It could degrade the localization ability of CB leading to errors and similar performance with DSB, as well as the signal reconstruction errors. Another possible reason for the similarity between DSB and CB in localization could be that only two sound sources are considered. CB would outperform DSB with the presence of many sources according to literature. The literature has been mainly focused on stationary sources, based on which CB delivers better localization. Whereas in the current study, moving instead of stationary sources have been addressed, and CB is found not advantageous over DSB in terms of localization. However, localization will not be further discussed due to the aim of signal reconstruction, not localization. Another finding is that e_{loc} is not correlated with e_{rec} according to the trend of the error curves.

The lowest e_{rec} with the window length of 32 samples and $\lambda = 1.75\beta$ can be selected for signal reconstruction. However, the corresponding e_{loc} reaches over 0.5 m in this sense, which would lead to perceptional difference in auralization. Additionally, this window length is too limited to extract the characteristics of the source signal, e.g. low frequency information. It is thus necessary to select parameters according to both localization and signal reconstruction.

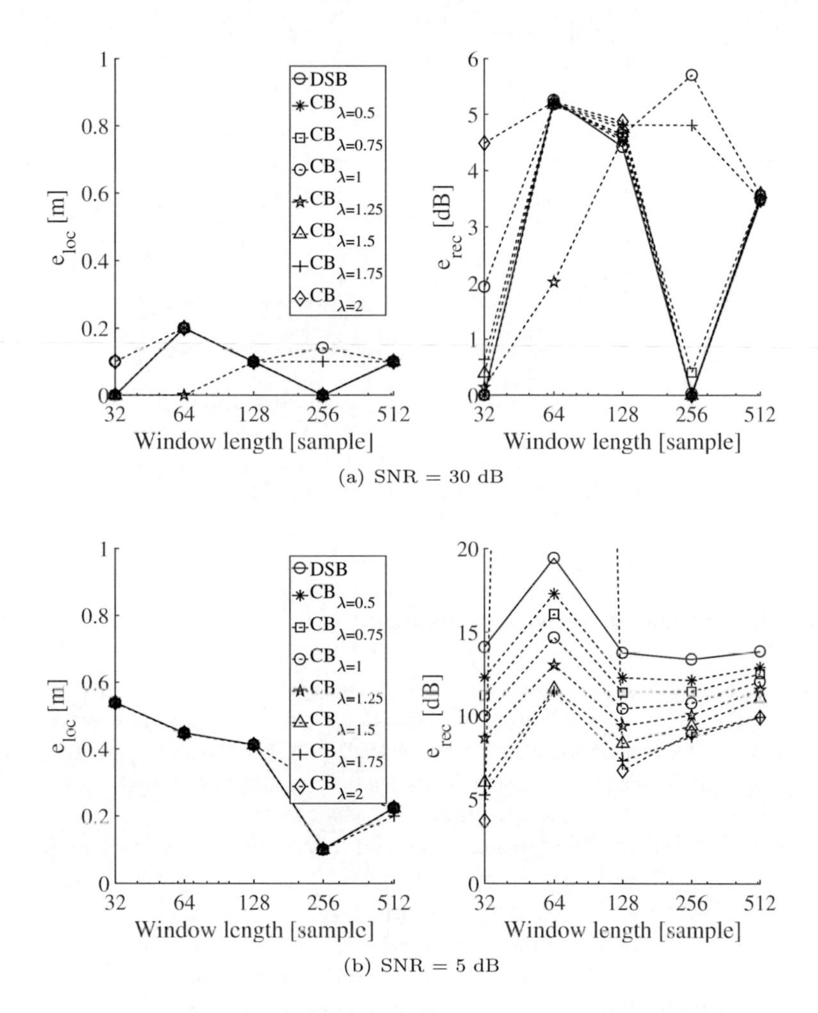

Figure 5.3.: The errors of localization and signal reconstruction versus window length for various regularization parameter λ with (a) SNR = 30 dB and (b) SNR = 5 dB using DSB and CB. The value of 64 samples with CB $\lambda = 2$ is 174.5 dB, which is out of the range of of the y-axis and not shown in the figure.

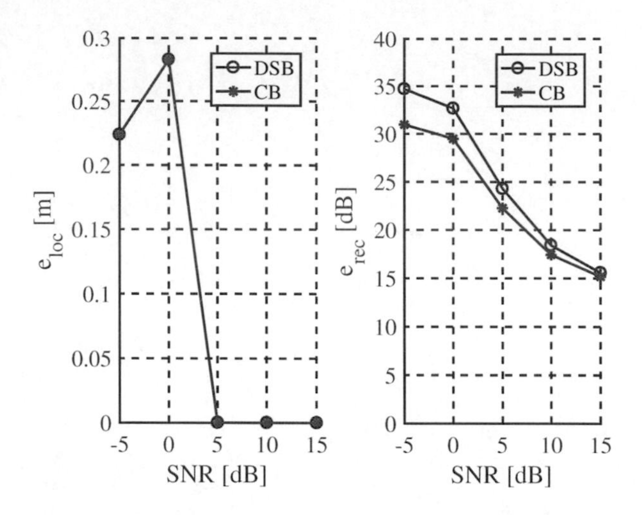

Figure 5.4.: The errors of localization and signal reconstruction versus SNR (Window length: 256 samples, $\lambda = 0.75\beta$).

5.2.2. Source speed vs. window length

The length of the viewing window, $t_{win} \times v$, determines the spatial extent, which would have an impact on localization using DSB [29]. Similarly, the combination of varying window length and source speed as in Section. 4 is investigated.

To attempt to a clearer illustration, a shaded area between the maximum and minimum e_{rec} with various source speeds, 20 m/s to 120 m/s with a step of 20 m/s, along the window length axis is given instead of an individual curve of each speed. In Fig. 5.5, the shaded areas of DSB and CB are given. Very large e_{mag} up to above 100 dB can be observed at the windows with 32 and 64 samples. As the window length grows, the CB errors are reduced significantly and are located below the DSB errors for the window length larger than 128 samples. Moreover, the error deviation difference between different speeds decreases with increasing the window length. The ℓ_2-norm of the source signal $\mathbf{s}$ in Eq. 2.30 would not be sufficient to represent the energy at the grid point on the reconstruction plane if the window is too short, because the samples might be small values in this short window. Therefore this analysis implies that a larger window length is preferable if CB is to be applied.

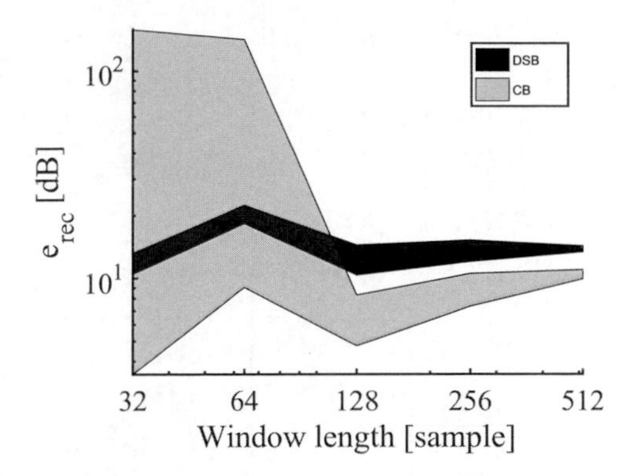

Figure 5.5.: The shaded areas represent the range of e_{rec} with varying source speeds from 20 m/s to 120 m/s with a step of 20 m/s along the window length axis using DSB and CB, respectively.

5.2.3. Mismatch

Mismatch emerges when a sound source is between two grid points. In the DSB case, wrong delays would be introduced to the calculations and basis mismatch in the sensing matrix would occur in CB [47]. In this context, neither beamforming method is able to correctly localize the source. The sensitivity of compressive sensing to DFT basis mismatch was studied by Chi et al. [88]. For the application of sound source localization using CB, the basis mismatch was analyzed and several wrong localization results were presented [48].

S1 is placed from 0.01 m to 0.09 m away from the origin of Ω in the y direction with 0.02 m step to create mismatch $\Delta \in [0.01 \text{ m}, 0.09 \text{ m}]$. SNR = 5 dB, $\lambda = 1.75\beta$ and the window length is 256 samples. The error results are exhibited in Fig. 5.6. e_{loc} using DSB and CB are identical, whereas the signal reconstruction using CB creates lower error than using DSB, the variation is around 5 dB. This is in line with the results from the previous section, that CB is more reliable than DSB under the given SNR if the parameters are selected properly. However, compared to the matched case, CB also yields larger e_{rec} due to mismatch compared to the matching cases.

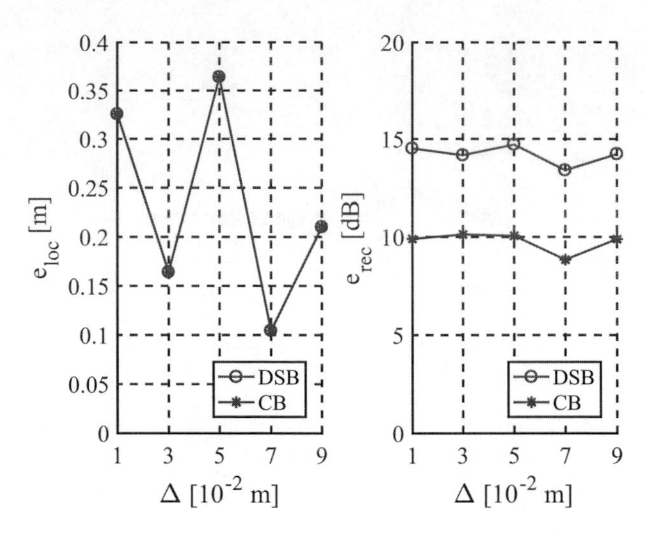

Figure 5.6.: The errors of localization and signal reconstruction versus mismatch Δ (Window length: 256 samples, SNR = 5 dB).

5.2.4. Distance

The positions of microphones have been randomized and optimized to reduce the coherence of the sensing matrix. Recalling Eq. 2.15, large $R(t)$ would increase the similarity between TDTFs in the sensing matrix, which could reduce the coherence. Together with the microphone position, the distance L between the source trajectory and array plane should also be considered with respect to the coherence.

Fig. 5.7 shows the drop of e_{rec} with decreasing L until 2 m, and the curve of CB is below that of DSB. When L =1 m, e_{rec} of CB rises and goes beyond DSB. This could be due to the regularization parameter λ. As L decreases the sensing matrix changes as well. λ was selected with $L = 5$ m, and it indicates that when L reaches 1 m λ is supposed to be reselected to balance the residual $||\mathbf{p} - \mathbf{Hs}||$ and the sparsity of $\mathbf{s}$.

However, in on-site measurements of pass-by vehicles, it is not always viable to place the microphone array close to the vehicle trajectory. First of all, the turbulence between the car and air would introduce more noise to the microphones. Additionally, the risk exists that the array aperture would fall towards the car.

It is thus necessary to keep the array distant from the trajectory. Note that in the safe distance range, CB outperforms DSB in signal reconstruction.

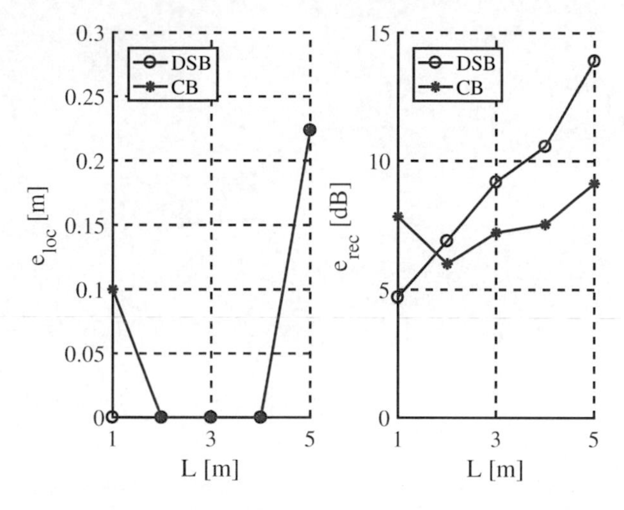

Figure 5.7.: The signal reconstruction error versus distance L (Window length: 256 samples, SNR = 5 dB).

5.3. Application on a moving loudspeaker

5.3.1. Measurement setup

The measurements were again performed on the Proving Ground of Institute for Automotive Engineering, RWTH Aachen University. This time the array was placed 5 m away from the moving trajectory of the near-side surface of the car to keep a safe distance. The loudspeaker was on the same position as in the measurements in Chapter. 4. During the measurements, the speeds were 20 km/h, 30 km/h, 50 km/h, 80 km/h and 100 km/h with two repetitions, respectively. Fig. 5.8 shows the pass-by measurement setup.

The same engine noise signal as in the simulations was played during the pass-by measurements. A sweep signal was added and played before the engine signal. The impulse response of the sweep signal could indicate the delay in the recording channels in terms of the playback channel and thus synchronize the recording

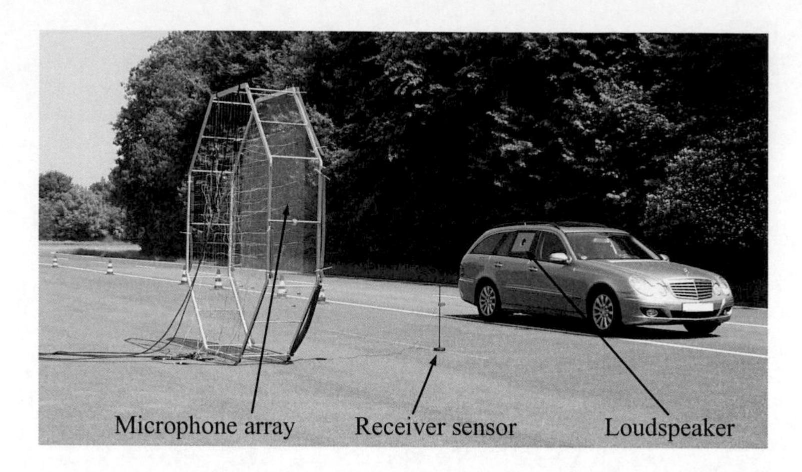

Figure 5.8.: The pass-by measurement of a car with a loudspeaker installed.

and playback, in order to extract the original signal from the playback channel according to the pass-by time $[T_{Apr.}, T_{Lea.}]$. The reference signal, from which the original signal was extracted, was recorded in an anechoic chamber.

The regularization parameter $\lambda = 1.75\beta$, and a steering window of 256 samples are applied.

5.3.2. Synchronization

To compare and evaluate the signal reconstruction, a reference signal is required. For the comparison in Chapter. 4, since the signal used is simply periodic, the original signal extracted from any time interval with identical length from the measured reference signal is the same. However, in the case of broadband engine noise, although it is considered as time-invariant, variations still exist between the original signals extracted from different small time intervals.

In the pass-by measurement, the loudspeaker and microphones were in two separate systems which were not synchronized. Therefore, in the following measurements, a sweep sound was first played by the loudspeaker to synchronize the recording channels of the microphones and the playback channels of the loudspeaker. After the sweep was played and recorded, the car drove away and had 20 s time to accelerate to a desired speed, as indicated in ig. 5.9(a). The equalized engine signals were subsequently played by the two channels of the loudspeaker. The playback signal is shown in Fig. 5.9(a). The channels of the photoelectric sensor and microphones were connected to RME OctaMic XTC, with a delay of 19 samples (Fig. 5.9(b) and (c)). The delay is then compensated. With deconvolution, the impulse response of the sweep can be localized in the recording channels as t_{IR} (Fig. 5.9(a)). From Fig. 5.9(b), the car pass-by time duration can be identified from the peaks and the duration is noted as T_c. Here the reference point for pass-by is the photoelectric sensor which was aligned with the array origin. The first peak indicates the car's first pass-by after playing the sweep and before playing the engine noise. Note that the playback was T_D delayed after the recordings started. The loudspeaker's location can be calculated through CB, so that time t_L the loudspeaker in front of the array origin can be found (Fig. 5.9(b)). Although the photoelectric sensor was synchronized with the microphones, the time it recorded was the emission time from the loudspeaker. Nevertheless, the recording and playback channels can be synchronized with the detection of t_{IR}. The temporal distance between the first sample of the engine noise signal and when the loudspeaker is in front of the array origin yields $T_1 = t_L - t_{IR} - 1.5 - 20$. Finally, the time stamp when the loudspeaker was in front of the array origin in the time axis of the reference signal is denoted as $t'_L = T_1 + L/c$, where $L = 5$ m is the distance between the loudspeaker and the microphone (Fig. 5.9(d)). The reference signal was measured in a hemi-anechoic chamber, where the microphone was placed on the ground to prevent the ground reflection.

With t'_L and the steering window length t_{win}, the original signal with the length $[t'_L, t'_L + t_{win}]$ can be extracted from the reference signal. Furthermore, the input signals for DSB and CB can also be obtained accordingly. The recording software is ITA-Toolbox, an open source MATLAB toolbox for acoustic measurements and signal processing [89].

5.3.3. Measurement results

The localization results of the loudspeaker in terms of various speeds are shown in Fig. 5.10. $[x', y']$ are the coordinates with the front bottom of Ω as the origin of the local coordinate system. The dashed lines are the approximated x' and y' positions of the geometrical center of the loudspeaker surface. However, it is uncertain if the geometrical center matches the acoustic center. An interesting observation is that CB is slightly more accurate in localizing the loudspeaker,

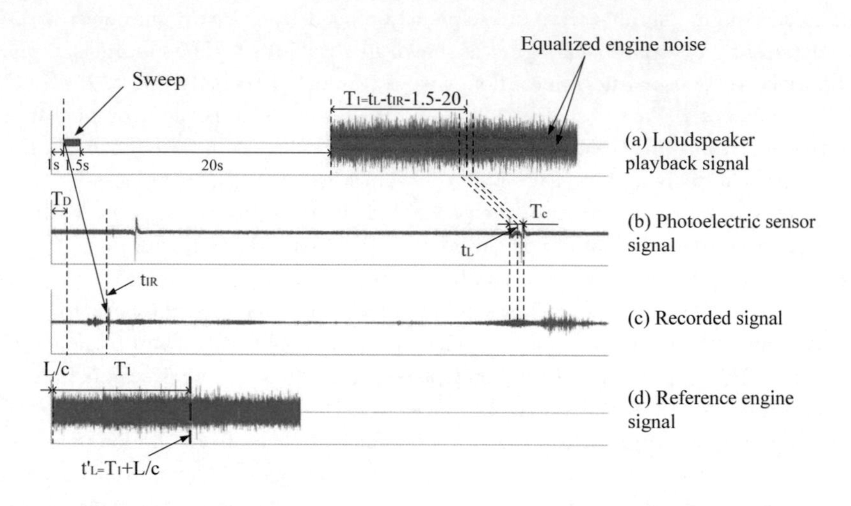

Figure 5.9.: Synchronization procedure between the playback and recording tracks during the pass-by measurements.

which was not implied in the simulations. However, it could result from the measurement uncertainties. The car could have not exactly followed the indicated line on the ground, especially at higher speeds. This uncertainty could also explain the large variations in the y' direction in Fig. 5.10. Moreover, uncertainties also exist in the measured positions, e.g. positions of the microphones, photoelectric sensors and loudspeaker, which can introduce errors into the results.

Take the first 50 km/h run as an example, the e_{rec} of DSB and CB are 4.4 dB and 3.5 dB, respectively. The errors are comparable with those in the simulations (e.g. Fig. 5.7 with $L = 5$ m, while e_{rec} of DSB in the measurements is even lower than in the simulation). However, the values might not be accurate since the aforementioned uncertainties could lead to incorrect distances and thus wrong time interval for the extraction of the original signal, on the basis of which e_{rec} is computed. To inspect the possible e_{rec} variations caused by uncertain time intervals, the calculated time interval is shifted from -256 to 256 samples, leading to 512 different extracted original signals. Fig. 5.11 shows the e_{rec} in terms of the sample shift, demonstrating that the e_{rec} bias ranges from 0 dB to 10 dB (for CB is around 0 - 9 dB).

Increasing the window length can reduce the e_{rec} bias in terms of sample shift, but it would be too computationally costly for CB. Especially in the measurement with unknown sound sources, a fine meshed reconstruction plane is desired and should be scanned completely. Requiring large computational resources is a limitation of CB. Nevertheless, the e_{rec} curve of CB is overall slightly below the DSB curve, which is in line with the simulations. Moreover, it also supports that CB is more robust under basis mismatch caused by the measurement uncertainties.

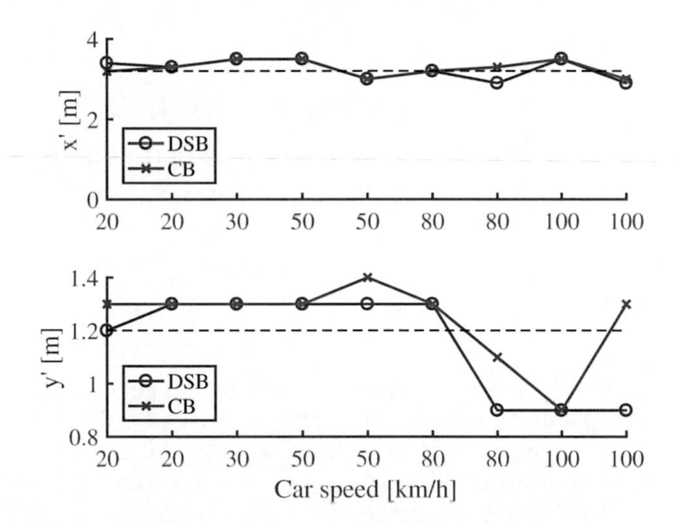

Figure 5.10.: Localization in x' and y' direction of the loudspeaker versus car speed with 0.1 m spaced grids. The dashed lines in the upper and bottom plots represent 3.2 m and 1.2 m, respectively. Here, $[x', y']$ are the coordinates with the front bottom of Ω as the origin of the local coordinate system.

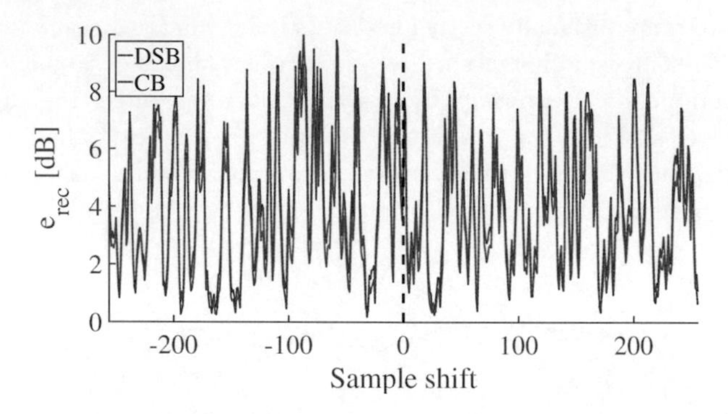

Figure 5.11.: The signal reconstruction error e_{rec} of DSB and CB in terms of sample shift of the time interval, which is used to extract the original signal.

5.4. Summary

CB is applied to model a moving engine signal using the pseudorandom microphone array. The parametric study and measurement application indicate that CB outperforms DSB in terms of signal reconstruction under noisy situations and basis mismatch, while under ideal noise condition, e.g. SNR = 30 dB, the two algorithms are quite similar. Additionally, the performance of CB varies in terms of the window length and distance L between the array and source moving trajectory. The reconstruction error increases with increasing L. CB shows more stable performance with larger window lengths for various source speeds. The 256-sample window is selected to provide better spectral resolution while not increasing e_{rec}. However, the window length is still limited to deliver robust signal reconstruction due to the uncertainties in the measurements. Larger window length will increase computational burden, which implies the limitation of CB. For localization, CB and DSB are quite similar in this study. Potential coherence in the sensing matrix due to large distance L and small number of sources could explain the localization similarity of the two algorithms. Nevertheless, for the purpose of signal reconstruction, CB has been demonstrated to be advantageous by means of simulations with various parameters and measurements.

6 Array measurement-based model for auralization

The main focus of this thesis is to model moving sound sources based on pass-by measurements by microphone arrays for auralization. In the previous chapters, the research has been conducted on array design and beamforming for localization and signal reconstruction for known moving sources. This chapter introduces a framework on synthesizing the moving sources (here two moving cars) for the auralization implementation in VR environments. First, pass-by measurements on two cars are presented with the designed pseudorandom array. Subsequently, the beamforming outputs are parameterized to represent the reconstructed signals, and parameter prediction is also introduced. Finally, how to incorporate the model in the VR system is introduced to provide dynamic moving sources for various scenarios.

6.1. Pass-by measurements

The model of moving sources utilizes the data from array-based pass-by measurements for the localization and signal reconstruction of moving sources. The pass-by measurements are considered as the data origin, as well as validation data for the synthesized source signals and future pass-by auralizations, as shown in the most right-hand box in Fig. 2.8.

Two cars were measured under the same measurement conditions in Chapter. 5 (Fig. 6.1). Both cars passed by the microphone array with constant speeds, ranging from 20 km/h to 120 km/h with 10 km/h step, with 5 m distance between the array and car trajectory. Each pass-by measurement for one specific speed was conducted two to three times to compensate for any potential failure that might happen in a singular measurement.

(a) Car01 pass-by measurement.

(b) Car02 pass-by measurement.

Figure 6.1.: Car pass-by measurements with the pseudorandom microphone array.

6.2. Localization

In the following contents, the middle steering window is applied with 256 samples. For the calculations using CB, $\lambda = 0.75\beta$. Two ways of mapping the source distribution are introduced as follows.

6.2.1. RMS map

Although a vehicle can be represented by a single equivalent point sound source [57], multiple sound sources are still desired to provide a more holistic representation of the main sound sources and thus deliver more realistic auralization. Therefore, the root mean square (RMS) values of each beamforming output calculated at each focus point on the reconstruction plane are used to detect the main sound sources of the pass-by cars.

For Car01 pass-by at 100 km/h, the localization maps using CB and DSB are shown in Fig. 6.2. As the emitted sound is mainly from the lower part of the car and due to the large computational cost using CB, only this part is processed. In the map of DSB, no clear sound source can be detected, while in the map of CB, four potential sources are given in the black circles in the top figure. Note that the advantage of CB over DSB in localization was not implied in the simulations in Chapter. 5, but similar observations can be found in the measurement results. Two remarkable sound sources can be identified in the front and back proximities

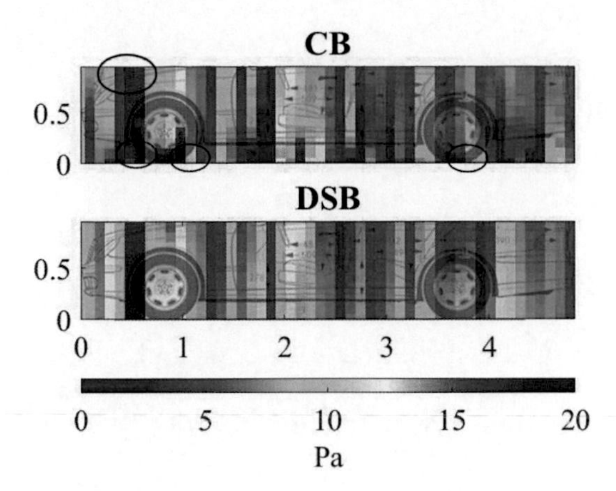

Figure 6.2.: The RMS localization maps using CB and DSB of Car01 pass-by at 100 km/h. The unit of the x and y axes is meter [m]. The black circles in the CB map represent potential sound sources.

of the front tire, and one slightly weaker source in the back of the rear tire. In addition, a possible sound source on the top of the car surface can also be observed, which could indicate the engine noise.

In this case, CB provides a clearer localization map than DSB. However, as the two maps are calculated in terms of the RMS values of the beamforming outputs, only the average energy distribution from the reconstruction plane is shown. Therefore, the resolution is not as high as the localization results using DSB reported in the literature, e.g. [32, 90] for cars or [33] for trains, which illustrated the localization results in frequency bands. In the following section, the localization is also be shown in frequency bands, i.e. 1/3 octave bands.

For Car02, Fig. 6.3 shows the localization maps using CB and DSB at the pass-by speed of 90 km/h. Similarly as Car01, CB outperforms DSB in localization with the RMS values. The results show that the noise emitted from the rear tire is located in the front, while only one potential source from the front tire is detected between the tire and car body. The circle on the top again implies the sound emitted from the engine.

The localization results shown by the RMS values are able to describe a rough source distribution on the car, whereas details are missing. However, the focus

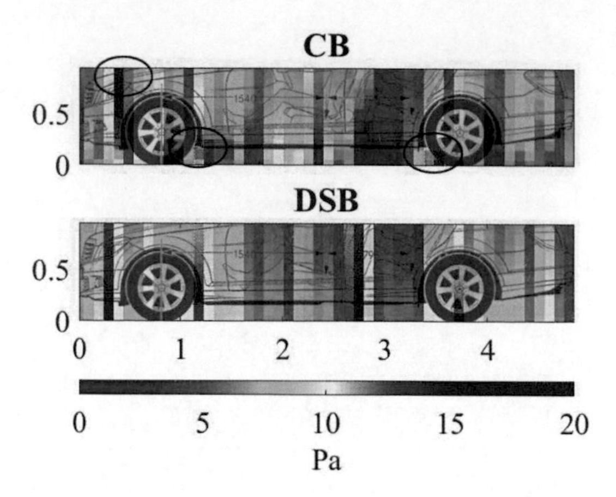

Figure 6.3.: The RMS localization maps using CB and DSB of Car02 pass-by at 90 km/h. The unit of the x and y axes is meter [m]. The black circles in the CB map represent potential sound sources.

of this thesis is not only on localization, but also on signal reconstruction. The detected sources can be regarded as equivalent sound sources. Using RMS values saves the computational cost and produces fewer sound sources compared with localization in frequency bands, wherein the source position could be frequency dependent [32]. Although objective evaluation criteria are able to take human hearing perception into account, listening tests are still required to evaluate the modeled moving sources and auralizations.

If the equivalent source method using RMS performs similar compared to the 1/3-octave-band method to be discussed later, the former method would be more beneficial due to the lower demand for calculation.

6.2.2. Third-octave-band map

Fig. 6.4 shows the localization maps of Car01 in 1/3 octave bands, i.e. 630 Hz, 800 Hz, 1 kHz, 1.25 kHz, 1.6 kHz, 2 kHz, 2.5 kHz and 4 kHz. As can be seen from the color bars, the energy decreases as the frequency increases, which indicates low-frequency sound dominates. It can be seen that the source locations are frequency dependent.

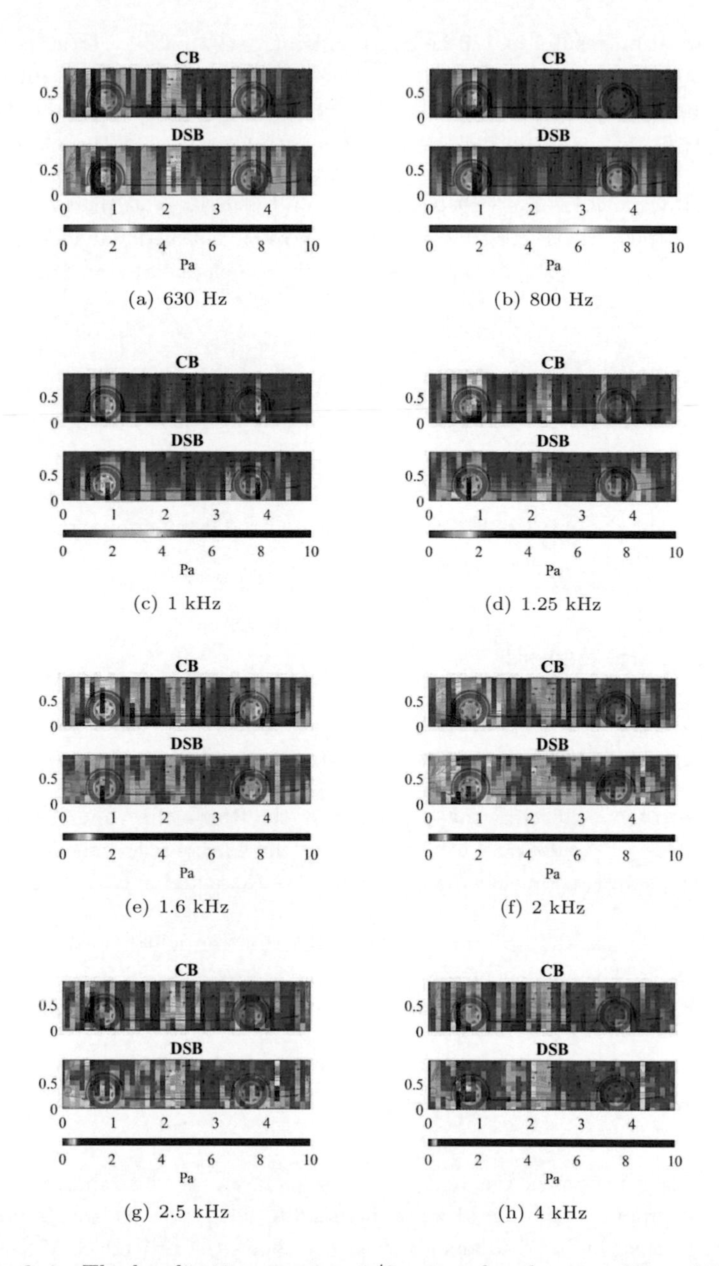

(a) 630 Hz (b) 800 Hz (c) 1 kHz (d) 1.25 kHz (e) 1.6 kHz (f) 2 kHz (g) 2.5 kHz (h) 4 kHz

Figure 6.4.: The localization maps in 1/3 octave bands using CB and DSB of Car01 pass-by at 100 km/h. The unit of the x and y axes is meter [m].

The localization results in 1/3 octave bands of Car02 pass-by at 90 km/h are given in Fig. 6.5. A sound source can be assumed at 2.8 m right on the ground according to Fig. 6.5(a), Fig. 6.5(f) and Fig. 6.5(h). This could be due to wind turbulence caused by the interaction between part of the car and the air. However, artificial effect caused by the algorithms could also explain this potential source. Additionally, the tires are still the main sound sources, and the locations are frequency dependent. For Car02, the engine noise is not as dominant as that for Car01 since no indication of source is shown in the engine area in Fig. 6.5.

6.3. Source synthesis

With the localization information, the beamforming outputs calculated at the focus points are regarded as the reconstructed signals of the corresponding sources. Parameters can be extracted from the reconstructed signals for signal synthesis and parameter prediction.

6.3.1. Spectral analysis

Equivalent sources

According to the localization maps in terms of the RMS values and Fig. 6.2, four sound sources are localized and shown as what the black circles indicate. The CB output signals of the four localized positions are extracted and given in Fig. 6.6.

The RMS map delivers an average energy distribution which shows equivalent sound sources. The equivalent sources may not reflect the real locations, leading to further study on the localization results in terms of 1/3 octave bands.

Multiple frequency-dependent sources

As the energy decays with increasing the frequency, only the bands with center frequencies from 630 Hz to 4 kHz are included to determine the source positions while neglecting higher frequencies. Fig. 6.7 shows the procedure of searching sound sources through the frequency bands. The indices of the localized sources of each band are given in Tab. 6.1.

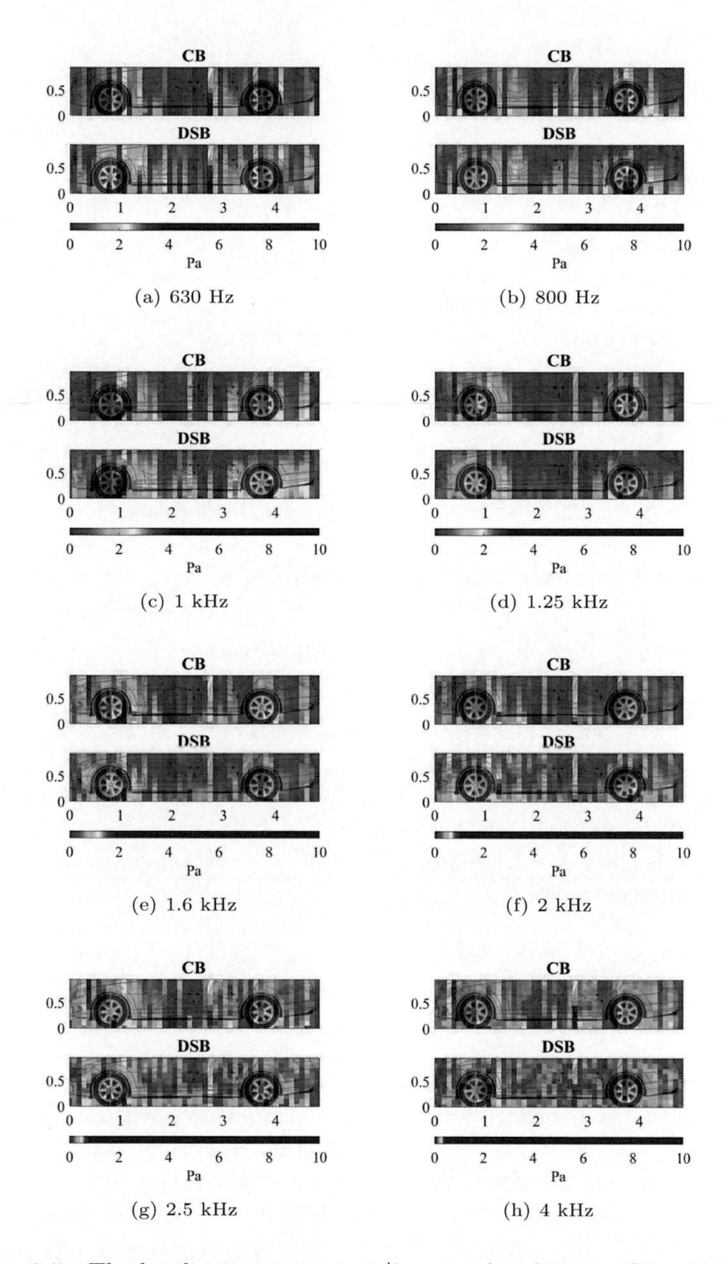

Figure 6.5.: The localization maps in 1/3 octave bands using CB and DSB of Car02 pass-by at 90 km/h. The unit of the x and y axes is meter [m].

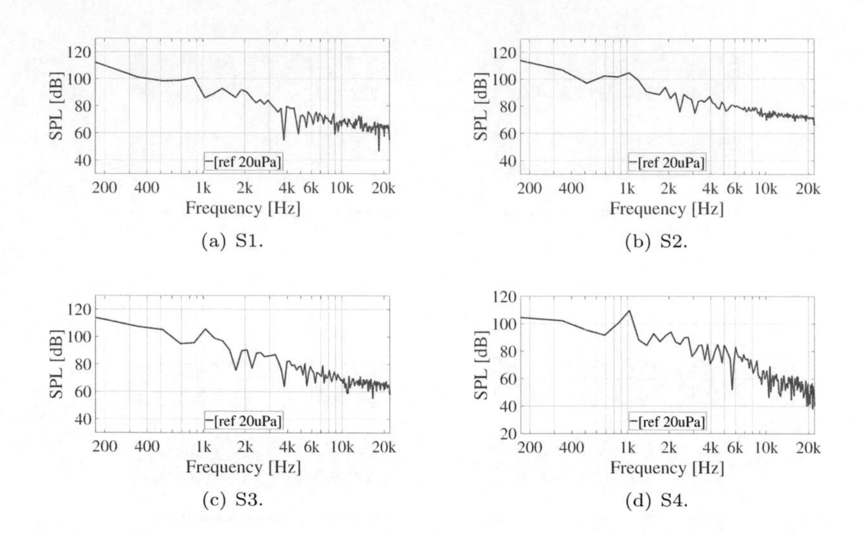

(a) S1.

(b) S2.

(c) S3.

(d) S4.

Figure 6.6.: Spectra of the CB outputs according to the positions of the four localized sound sources in terms of the localization maps in 1/3 octave bands.

From Tab. 6.1 it can be found out that some sources are broadband due to their repetitive occurrences in various bands, e.g. the coordinates of (7 1) and (10 1), which indicate the sources from the front tire. For the rear tire it is clearer since only one source position is detected from all the frequency bands. Additionally, all the other sources are in the frontal part of the car, mainly around the front tire and engine part.

Fig. 6.8 illustrates the localized sound sources according to the results of the 1/3 octave bands with center frequencies from 630 Hz to 4 kHz and the spectra of the corresponding beamforming outputs, i.e. the reconstructed signals of the sources.

However, for auralization, arbitrary lengths of signals are desired to provide dynamic virtual scenarios with flexibilities, which makes the 256-sample signals (sampling rate 44100 Hz) not sufficient. Another concern is that a short-time signal lacks of spectral information, which can result in inappropriate spectral analysis and synthesis. Therefore, the signal length needs to be extended to fulfill the requirement. Moreover, the reconstructed signals can only represent the scenarios where the measurements take place, but not unknown scenarios. For instance, there is no knowledge about the localization and signal reconstruction of unknown speeds which are not included in the measurements, e.g. pass-by at 130 km/h or larger speeds. To overcome the restrictions mentioned above, parameters need to be extrapolated from the reconstructed signals and proper prediction methods need to be adopted to predict unknown scenarios.

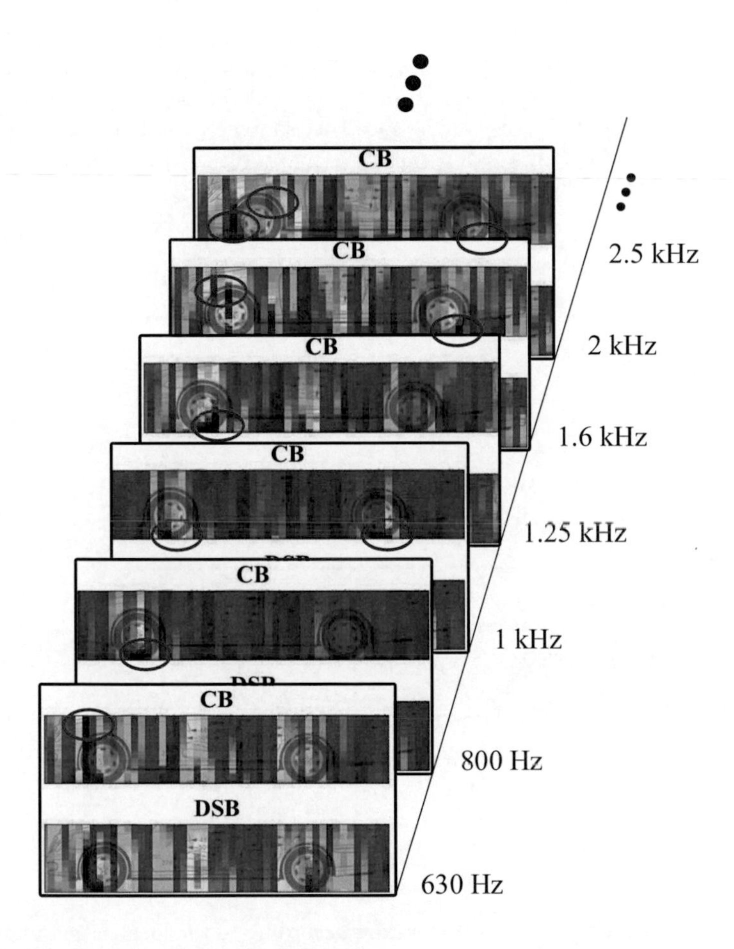

Figure 6.7.: The localized source positions in the localization maps in 1/3 octave bands. The red circles represent potential sound sources (Car01).

0.63	0.8	1	1.25	1.6	2	2.5	3.15	4
(7 2)	(10 1)	(10 1)	(7 1)	(9 6)	(6 4)	(7 6)	(7 8)	(7 3)
(7 10)		(40 1)	(10 1)	(15 3)	(40 1)	(40 1)	(11 1)	(40 1)
				(7 8)		(11 1)	(40 1)	

Table 6.1.: The source indices in each 1/3 octave band in the local coordinate system (Car01). The numbers in the first row are the center frequencies of the bands (unit: kHz), and the second row shows the coordinate indices of the localized sources (The left bottom point on the reconstruction plane is the origin of the local coordinate system).

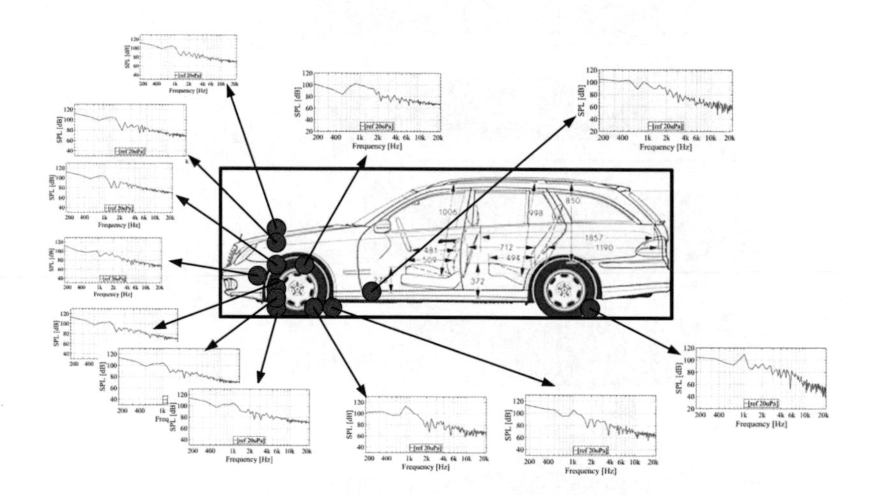

Figure 6.8.: The localized source positions according to the localization maps in 1/3 octave bands and the spectra of the corresponding reconstructed signals.

6.3.2. Parameterization

The tire noise gradually dominates as the speed increases. The previous examples showed no clear localization of the engine noise. On the one hand, the speed selected is large so that the engine noise was lower than the tire noise; on the other hand, the window length was small so it can also lead to the omission of weaker sound sources, especially at low frequencies. According to Fig. 4.8 in Chapter. 4, more accurate results can be yielded at lower speeds with larger window lengths. Therefore, a larger window size (8192 samples) and lower speed (30 km/h) is selected for the analysis of Car01.

In the localization map of the 1/3 octave band with the center frequency of 1.6 kHz (Fig. 6.9), the dark color area which implies the engine noise source can be found on the left top. After extracting the DSB output signal and applying the peak detection method [63], the spectrum with detected peaks is shown in Fig. 6.10. The detected frequencies are, 64.6 Hz, 123.8 Hz and 236.9 Hz, respectively, which can be the order frequencies of the engine. Higher order frequency components are unable to be detected from the spectrum. In this sense, following the idea of Pieren et al. [16], the higher order tones can be synthesized based on the detected tones (Another reference is [19], which applies the proposed method of Pieren et al and synthesized car engine noise).

Following the SMS method [63], the residual and envelope of the residual are calculated, as shown in Fig. 6.11.

For tire noise, since no tonal components are detected, the envelope from the reconstructed signal can be directly extracted without spectral subtraction.

6.3.3. Spectral synthesis

With SMS, a reconstructed signal can be decomposed into tonal and broadband components. For the tonal component, the parameters for signal synthesis consists of the amplitude A, frequency f and phase θ (if necessary), and FFT coefficients (in the case Fourier transform is used as the time-frequency transformation) are the synthesis parameters for the broadband component. White noise is filtered with the envelopes of the residuals, which are as the filters during the synthesis procedure.

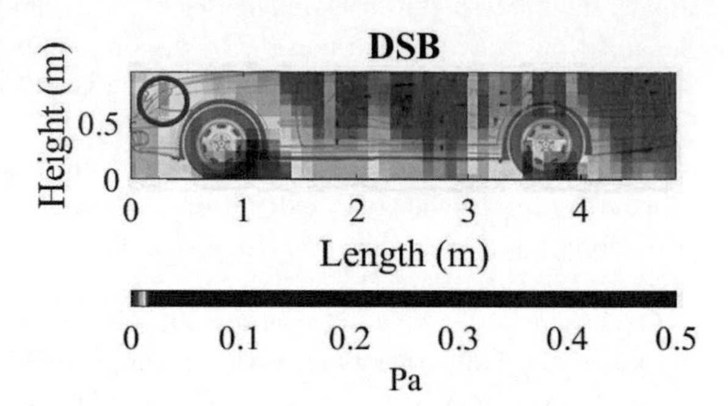

Figure 6.9.: The localization map of the 1/3 octave band with center frequency of 1.6 kHz. Window length: 8192 samples, car speed: 30 km/h. The black circle indicates the potential engine noise source.

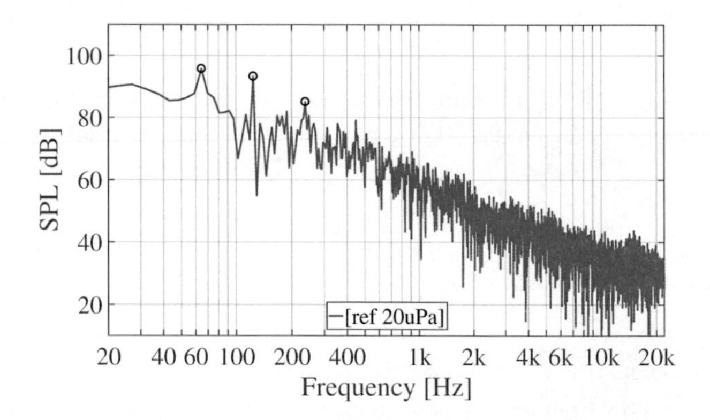

Figure 6.10.: The three detected peaks shown on the spectrum of the beamforming output spectrum in terms of the localization result in Fig. 6.9. Window length: 8192 samples, car speed: 30 km/h.

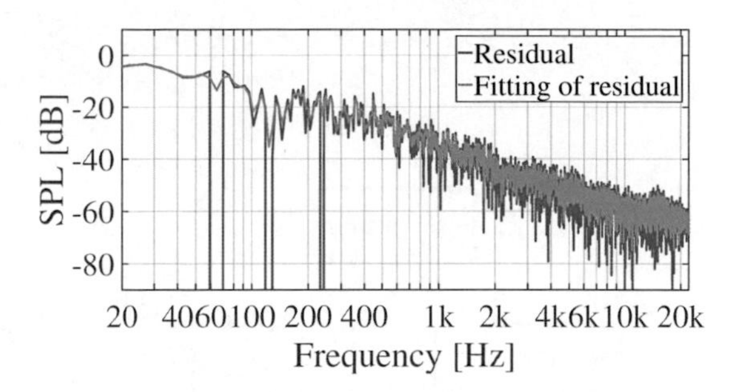

Figure 6.11.: The residual and residual envelope of the detected engine noise.

6.3.4. Parameter prediction

As mentioned in the introduction, auralization would be not sufficient if only the measured scenarios are applied. Take the source speed as an example for parameter prediction, as the pass-by measurements are not able to cover all speeds, it is necessary to investigate prediction methods to predict the synthesis parameters of under the conditions of unknown speeds.

A model to predict the tonal component of the engine noise in terms of the car speed can be found in Pieren et al. [16]. The model can be used to predict the tonal parameters based on the extracted parameters from the reconstructed signals. For the broadband component, the relationship between the level of the broadband component and the speed can be predicted. However, how the spectral envelope varies in terms of the speed still remains an open question. Non-linear prediction models could be investigated for the prediction of the FFT coefficients. Although the directivity may be not relevant in the source modeling, in large scale auralization, e.g. long motion trajectory, the directivity becomes more significant to deliver realistic acoustic perception [91, 7, 10].

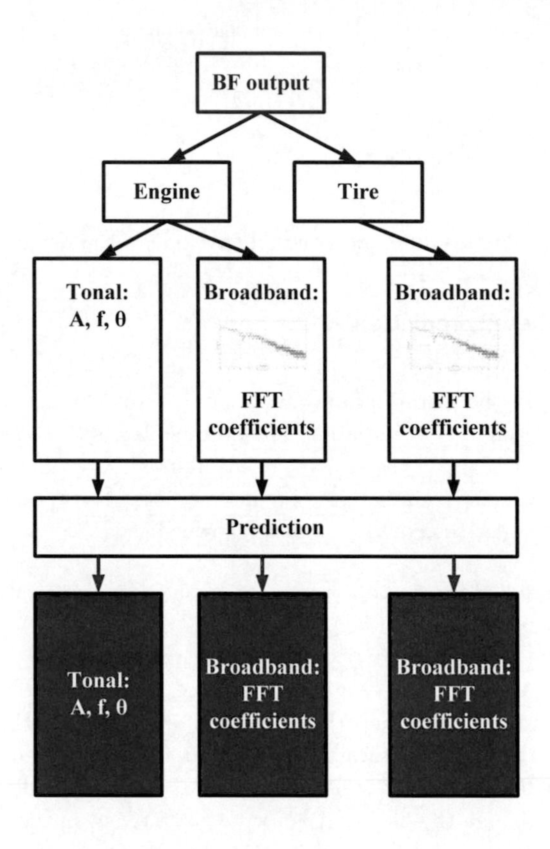

Figure 6.12.: Parameter prediction based on the parameters extracted from the beamforming outputs, i.e. the reconstructed signals from a car. The tonal component from the engine can be synthesized and predicted by the amplitude A, frequency f and phase θ, and the broadband components from the engine and tire can be synthesized and predicted by the FFT coefficients.

6.4. Incorporation in virtual reality systems

The signal, position and directivity are the three main characteristics of the source modeling for auralization in VR [25]. In addition, the speed, motion orientation, acceleration etc. are parameters that should also be taken into consideration when auralizing moving sources in VR. This thesis aims at obtaining the signal and position as in the main characteristics, including the parameters of speed and motion orientation of the moving sound source. However, the directivity pattern which is important, yet often forgotten [92], is not discussed. As the chosen steering window length is small and the source speed is not very large, the directivity pattern varies little during the motion of the source and thus is considered as constant.

6.4.1. Directivity

To incorporate the directivity in the source model in the VR system, existing directivity patterns and measurements can be adopted. The directivity of a point source can be decomposed into horizontal and vertical directivities [93]. For a car, the directivity of tire or engine noise source is delivered as

$$D(f,\theta,\psi) = D_H(f,\theta) + D_H(f,\psi) \tag{6.1}$$

where θ is the horizontal angle as in Fig. 2.1, ψ is the vertical angle, f is the frequency and D_H is the horizontal directivity. D_H is expressed in terms of the height of the source. At the height of 0.01m,

$$D_H(\theta) = \begin{cases} 0, & \text{f} < 1250\text{Hz} \\ (-3+5|\cos(\theta)|)\sqrt{\cos(\psi)}, & \text{f} \geq 1600\text{Hz} \end{cases} \tag{6.2}$$

$$D_H(\theta) = (1.546(\pi/2 - \theta)). \tag{6.3}$$

At the height of 0.75m,

$$D_H(\theta) = (1.546(\pi/2-\theta)^3 - 1.425(\pi/2-\theta)^2 + 0.22/\pi/2-\theta+0.6)\sqrt{cos(\psi)}. \quad (6.4)$$

The directivity can also be obtained by measurements. The directivity measurement of railway noise sources has been proposed by Zhang [94]. However, the directivity of the noise source caused by the source motion, is still difficult to measure. In their work only empirical models are given. Directivity measurement requires a spatially extended microphone distribution to cover a large range of angles. This requirement constrains the microphone arrays proposed in this thesis to conduct the measurements.

6.4.2. Virtual Acoustics

To achieve auralization in VR, a complete source model should be combined with proper propagation models and reproduction techniques. The real-time VR platform, Virtual Acoustics (VA) [95], recently launched by the Institute of Technical Acoustics (ITA), RWTH Aachen University can incorporate the parameters of the sources, propagation models, and create dynamic and interactive sound scenes [96], including moving sources. The platform allows real-time interactions between the users and moving sources by controlling the input parameters. In addition, VA can also be jointly used with the visual VR system [97].

Therefore, for the further research based on the model proposed in this thesis, e.g. pass-by auralizations to validate the signal reconstruction through listening tests can be performed in the VA platform. Auralizing moving vehicles in urban environments can be achieved by combining with the visual part for the purpose of noise management and urban planning.

6.5. Summary

This chapter provides a framework on modeling moving sound sources based on array measurements for the auralization in VR. Two pass-by cars were measured with the pseudorandom array. Equivalent and frequency-dependent sources are identified from the RMS and 1/3-octave-band localization maps, respectively. With the array steering the localized positions, the signals are reconstructed from the beamforming outputs which potentially represent the sound sources on the cars. However, it is not clear how much difference the auralizations would have in terms of the localization and signal reconstruction from the RMS and 1/3-octave-band localization maps. Parameterization is subsequently conducted and parameters representing the tonal and broadband components are extracted. With the parameters, the source signals can be synthesized. The parameters would also benefit obtaining parameters of unknown measurement scenarios. Combining with directivities, a complete model for moving sound sources will be established. In this thesis, parameter prediction, signal synthesis and directivity are briefly introduced but not elaborately studied. In the end, VA is introduced as a moving source auralization platform for future research.

7

Conclusion and outlook

In this thesis, a model of moving sound sources has been developed based on microphone array measurements. The model extracts the spatial locations and signals for the purpose of auralizing moving sources. Two beamforming algorithms, i.e. DSB and CB, are considered as the main algorithms in the model. The following contents are addressed in this work:

- A spiral and a pseudorandom microphone array are designed for DSB and CB, not only for localization, but also for signal reconstruction;
- DSB and CB are applied for localization and extended for signal reconstruction to model moving sound sources;
- guidelines for using the array measurement-based model are investigated, and a framework of using this model for auralization is provided.

The thesis consists of four main parts. First of all, microphone arrays are briefly introduced and a spiral and a pseudorandom array are designed for DSB and CB in Chapter. 3. In Chapter. 4, DSB is applied to localize and reconstruct a moving periodic sound source. Furthermore, a broadband signal, a moving engine signal is localized and reconstructed using CB in Chapter. 5. In the two chapters, parameters in the model are discussed. Finally, a synthesis framework of the measurement-based model is exploited in Chapter. 6.

In Chapter. 3, the design of a spiral array and a pseudorandom array is employed. A spiral array is designed for DSB. A pseudorandom array is designed to meet the RIP requirement of CB, and to assure the localization performance of DSB as well by constraining resolution and MSL. The array configuration with 19 dB MSL and 15° resolution at 30° steering angle is finally determined as the optimal configuration for the sake of both CB and DSB. The two arrays are subsequently constructed and applied in on-site measurements.

With the two designed arrays, moving sound sources are modeled in terms of localization and signal reconstruction using DSB and CB. In Chapter. 4, DSB is modified for moving sources, the localization and signal reconstruction results are evaluated by studying the errors of localization, frequency and level by varying the parameters included in the model, i.e. steering position, steering window length and source speed. The middle position of the steering window is adopted in this thesis due to its lower errors than the left and right windows. It is found that the steering window and the source speed have significant influence on the localization and reconstruction accuracy of the model. Additionally, the localization accuracy could be degraded by large windows and low frequencies, spectral leakage and errors caused by de-Dopplerization. For signal reconstruction, the frequency reconstruction has negative correlation with the window length and source speed. The level errors show similar varying tendency as the localization errors in the x direction. Moreover, the errors appear to positively correlate with the source speed and window length at higher speeds. Various window lengths are necessary to be applied in terms of various source speeds to deliver more accurate results. The combination of limited frequency and spatial resolution, spectral leakage and errors caused by de-Dopplerization would result in unexpected deviations. This reason also holds for the errors using CB in Chapter. 5. The application of the model using DSB on a pass-by loudspeaker shows the capability of localization, and acceptable deviations of signal reconstruction in terms of frequencies and levels of the periodic signal.

A broadband engine signal which consists of periodic and broadband components is explored using CB. The results are shown together with the results using DSB in Chapter. 5. For localization, CB and DSB are quite similar in this study. Potential coherence in the sensing matrix due to large distance L and small number of sources could explain the localization similarity of the two algorithms. Moreover, the parametric study and measurement application indicate that CB outperforms DSB in terms of signal reconstruction under noisy situations (SNR = 5 dB) and basis mismatch, while under ideal noise condition, e.g. SNR= 30 dB, the two algorithms are quite similar. In addition, the performance of CB varies in terms of the window length and distance L between the array and source moving trajectory. The reconstruction error increases with increasing L. With larger window lengths, CB shows more stable performance in terms of various source speeds. The 256-sample window is selected to provide better spectral resolution while not increasing e_{rec}. However, the window length is still limited to deliver robust signal reconstruction due to the uncertainties in the measurements. In the measurements, although uncertainties influence the results, CB still outperforms DSB regarding signal reconstruction. Larger window length will increase computational burden, which implies the limitation of CB.

Finally, a framework of the array measurement-based synthesis for auralization is proposed in Chapter. 6. The model using DSB and CB is applied on two pass-by cars. Equivalent and frequency-dependent sources are identified from the RMS and 1/3-octave-band localization maps. The signals extracted according to the localized sources are parameterized in tonal and broadband components following the idea of SMS [63]. Parameters from the decomposed components, i.e. level (amplitude) A, frequency f and phase θ (if necessary) for the tonal component and FFT coefficients for the residual can be used for signal synthesis, and more importantly, for the prediction of the parameters in the scenarios that are not included in the pass-by measurements. Combining with the directivity information of the sources, the source modeling is complete for auralization. However, covering the signal synthesis, prediction and directivity achievement would exceed the scope and time schedule of this thesis. Therefore, only the ideas are included under the framework, and no elaborated investigation is performed.

The proposed model is capable to provide the locations and signals of moving sound sources, more specifically, a moving periodic signal and an engine signal. Therefore, the conclusions are drawn on the basis of the two given cases. More various signals need to be investigated with the proposed model. In addition, the model is insufficient to synthesize impulsive sounds, e.g. squeal and rattling sounds from trains. An impulsive sound can be reconstructed separately by adjusting the steering window length according to the impulsive time duration, and extracting the time-domain impulsive signal just from the beamforming output. Since the occurrence of the impulsive sound is intermittent, the localization and signal reconstruction might be degraded due to large steering angle if the sound does not happen right in front of the array.

Furthermore, CB is more promising to provide more accurate localization according to the stable source research [44, 45, 46, 47, 48]. However, in this thesis, the localization abilities of the two algorithms are quite similar. Only in the measurements, CB is more advantageous than DSB. Further work on increasing the localization ability of CB which might have a positive influence on the signal reconstruction is necessary. For example, designing arrays which are able to provide less-coherent sensing matrices, and higher order sparsity needs to be exploited for identifying multiple moving sound sources.

Moreover, more efficient objective evaluation methods are required to better map human perception, e.g. psychoacoustic parameters [14]. The measure e_{rec} proposed in this study is advantageous to quickly evaluate the reconstructed signals. However, although e_{rec} accounts for human perception, listening tests are still desired to acquire more straightforward and accurate perceptual differences between the reconstructed and original signals.

Last but not least, combining with proper propagation models and reproduction techniques, pass-by auralizations can be created to compare with real-scenarios to help validate and improve the proposed model.

Acknowledgments

First of all, I would like to thank Prof. Michael Vorländer for letting me in his group and doing research in acoustics, which is so different from other subjects for the opportunity not only to see our research, but more importantly to listen to it. He has been inspiring, helping and guiding me through the PhD years with his wisdom of being a scholar and a person. I also appreciate Prof. Peter Jax's will to be my second supervisor.

My gratitude goes to Prof. Diemer de Vries, the former guest professor at ITA. Without him, I wouldn't be in this institute. His help and responsibility for me during my first years gave me so much inspiration and motivation.

I would like to thank Karin Charlier for showing up every time when I encountered any kind of trouble in life and at work. Thanks to Uwe Schlömer and colleagues from the mechanical workshop for realizing the microphone arrays with the delicate work, and to Rolf Kaldenbach and Norbert Konkol from the electronic workshop for all their contributions to my measurements.

Special thanks go to Dr. Gottfried Behler for many fruitful talks regarding research and measurements. He is full of practical ideas to transform theoretical design to feasible plans. I am thankful for him and his car for working on the tedious outdoor pass-by measurements in hot summer days. The passion, enthusiasm and knowledge from Dr. Frank Wefers had been leading me to the right direction in the first two years. I would like to express my gratitude for Johannes Klein and Marco Berzborn for the talks on my measurements. Thanks to Dr. Markus Müller-Trapet for his MATLAB codes to get me familiar with beamforming more quickly, and the VR-crew: Lukas Aspöck, Jens Mecking, Michael Kohnen, Jonas Stienen and Muhammad Imran.

My ITA memory wouldn't be complete without sharing the years' time with others. Special thanks go to Dr. Noam Shabtai as my officemate and friend from my first days at ITA. His humor and jokes made a wonderful atmosphere in the office. Thanks go to Dr. Wanglin Lin, Margret Engel, Dr. Samira Mohamady, Rob Opdam, Dr. Ellen Peng and Dr. Shuai Lyu for all those fun we had after work. I would like to thank my students, namely Yan Li, Wenhuan Duan, Long Chen, Xinshuo Gu for contributing to this work.

Thank every ITA colleague for being my voluntary German teachers. Their patience and willingness helped improve my German skill a lot. Thanks also go to other colleagues, Dr. Martin Guski, Ingo Witew, Dr. Martin Pollow, Rhoddy Viveros, Florian Pausch, Mark Müller-Giebeler, Dr. Ramona Bomhardt and Josefa oberem.

I would like to thank my parents for all the support through the years, letting me be the person I wanted to be. Last but above all, I would like to thank my beloved wife, Bo Zhang, for the lasting love, confidence and belief in me. She is always the person behind me and backs me up by any means.

Virtual recordings of moving sound sources

In the measurement with microphones, the microphones are usually connected to the same sound card and thus synchronized. Therefore, the recording length of every microphone is the same. In the MATLAB code below, zeros are added to the calculated recorded signals to make all the recordings with same length. In addition to recording moving sound sources, the case of stationary sound source is also accounted for.

```
function [sig_R_final] = rec_movingSrc( sig_S, pos_s,
    pos_r)
%REC_MOVINGSRC Simulates the microphone recordings of
    moving sound sources
%   This function can return recorded signals of
    microphones from multi
%   moving sound sources
%
%   Syntax: out = rec_movingSrc( sig_S, pos_s, pos_r)
%   sig_S: source signals
%   pos_s: initial positions of the sources
%   pos_r: positions of microphones
%   c:     speed of sound
%   v_s:   source speed
%   ori:   moving orientation of the sources
%   fs:    sampling rate
%   tRun:  motion time (= sound emission time)

%   Reference <Theoretical Acoustics>
%   (11.2: Source emission from moving sources)
```

```
global c v_s ori fs tRun
[L,nS] = size(sig_S);
[~,nR] = size(pos_r);
timeResolution = 1/fs;
M = v_s/c;

if L/fs < tRun
    error('Source signal not long enough for the input
        run time.');
end
if norm(sourceOrientation) ~= 1
    error('Source orientation should be normalized.');
end

if v_s < 0
    error('The speed of source cannot be minus!');
elseif v_s > 0
    speedVector = ori * v_s;
    runSample = ceil(fs * tRun);                  % ceil: to
        the upper integer
    sampVect_e = 1:runSample;
    tVect_e = (sampVect_e - 1)' * timeResolution;
    distSRAll = zeros(nR,nS,runSample);
    cosTheta = zeros(nR,nS,runSample);
    for iR = 1:nR
        sumSig_r{iR} = [];
        sumTemp = [];
        for iS = 1:nS
            if iR == 1
            % calculate only once. sources are not
                dependent on receivers
                sourcePositionAll{iS} = repmat(pos_s(:,
                    iS),[1,runSample])...
             + repmat(speedVector,[1,runSample]) .*
                 repmat(tVect_e',[3,1]);
            end
            % distSRAll is R in the book
            distSRAll(iR,iS,:) = sqrt(sum((
                sourcePositionAll{iS} ...
```

```
        - repmat(pos_r(:,iR),[1,runSample])).^2)
            );
    vect_r2s = repmat(pos_r(:,iR),[1,runSample])
        - ...
        sourcePositionAll{iS};
    cosTheta(iR,iS,:) = vect_r2s' * ori...
        ./ norm(vect_r2s);
    sig_r_unEq{iR,iS} = ...
        1/4/pi./squeeze(distSRAll(iR,iS,:)) ./
            (1 - M * ...
        squeeze(cosTheta(iR,iS,:))).^2 ...
        .* (sig_S(:,iS));
    % interpolation in reception time. Convert
       signals from unequal
    % time interval to equal
    tTrans{iR,iS} = squeeze(distSRAll(iR,iS,:))
        ./ c;
    tVect_r_unEq{iR,iS} = tVect_e + tTrans{iR,iS
        };
    tVect_r_Eq{iR,iS} = ...
        (tVect_r_unEq{iR,iS}(1):timeResolution:
            ...
        tVect_r_unEq{iR,iS}(end))';
    sig_r_Eq{iR,iS} = ...
        interp1(tVect_r_unEq{iR,iS},...
        sig_r_unEq{iR,iS},...
        tVect_r_Eq{iR,iS},'spline');
    % complete the signal in global time with
       zero paddings before
    % the first sample arrives at the receiver
    tTrans_1Samp(iR,iS) = distSRAll(iR,iS,1)/c;
    nZrs(iR,iS) = ceil(tTrans_1Samp(iR,iS) * fs)
        ;
    sig_r_zrs{iR,iS} = ...
        [zeros(nZrs(iR,iS),1);sig_r_Eq{iR,iS}];
    len_sig_r_zrs(iR,iS) = ...
        numel(sig_r_zrs{iR,iS});
    % compensate all signals into same length in
        one microphone
    if nS == 1
```

```
                sumTemp = sig_r_zrs{iR,iS};
            else
                if iS == 1
                    sumTemp = sig_r_zrs{iR,iS};
                else
                if len_sig_r_zrs(iR,iS) >= length(
                    sumTemp)
                lengthDiff = len_sig_r_zrs(iR,iS) -
                    length(sumTemp);
                sumTemp = sig_r_zrs{iR,iS} + [sumTemp;
                    zeros(lengthDiff,1)];
                else
                lengthDiff = length(sumTemp) -
                    len_sig_r_zrs(iR,iS);
                sumTemp = sumTemp + [sig_r_zrs{iR,iS};
                    zeros(lengthDiff,1)];
                end
                end
            end
        end
        sumSig_r{iR} = sumTemp;
        lengthReceiver(iR) = numel(sumSig_r{iR});
    end
    maxLengthReceiver = max(lengthReceiver);
    % compensate signals in all microphones with same
       length
    for iR = 1:nR
        lengthDiffReceiver = maxLengthReceiver -
            lengthReceiver(iR);
        sig_R_final(:,iR) = [sumSig_r{iR};...
            zeros(lengthDiffReceiver,1)];
    end

    % v_s = 0
else
    for iR = 1:nR
        distSR(iR,:) = sqrt(sum((pos_r(:,iR) - pos_s)
            .^2));
        t_trans(iR,:) = distSR(iR,:)/c;
        samp_trans(iR,:) = round(t_trans(iR,:)*fs);
```

```
        for iS = 1:nS
            sig_R_tShift{iR,iS} = [zeros(samp_trans(iR,
                iS),1);sig_S(:,iS)];
            sig_R_tShift_1r{iR,iS} = sig_R_tShift{iR,iS}
                / distSR(iR,iS);
            % sum all source signals at iR
            if iS == 1
                sumTemp = sig_R_tShift_1r{iR,iS};
            else
                if length(sig_R_tShift_1r{iR,iS}) >=
                    length(sumTemp)
                    sumTemp = sig_R_tShift_1r{iR,iS} +
                        ...
                        [sumTemp;zeros((length(
                            sig_R_tShift_1r{iR,iS})...
                        - length(sumTemp)),1)];
                else
                    sumTemp = ...
                        [sig_R_tShift_1r{iR,iS};zeros((
                            length(sumTemp)...
                        -length(sig_R_tShift_1r{iR,iS}))
                            ,1)] + sumTemp;
                end
            end
        end
        sig_R{iR} = sumTemp;
        len_sig_R(iR) = length(sig_R{iR});
    end
    % compensate signals in all microphones with same
        length
    maxLength = max(len_sig_R);
    for iR = 1:nR
        len_diff = maxLength - len_sig_R(iR);
        sig_R_final(:,iR) = [sig_R{iR};zeros(len_diff,1)
            ];
    end
end
end
```

Measurements of pass-by trains

Most results of this preliminary study were published in [70].

B.1. Array performance

A uniform linear array with 24 microphones was used to localize the pass-by trains.

The array beam patterns at 2 kHz with two different weighting are shown in Fig. B.1. It can be seen that using the Chebyshev spatial weighting increases the beamwidth but lowers the maximum side lobe level.

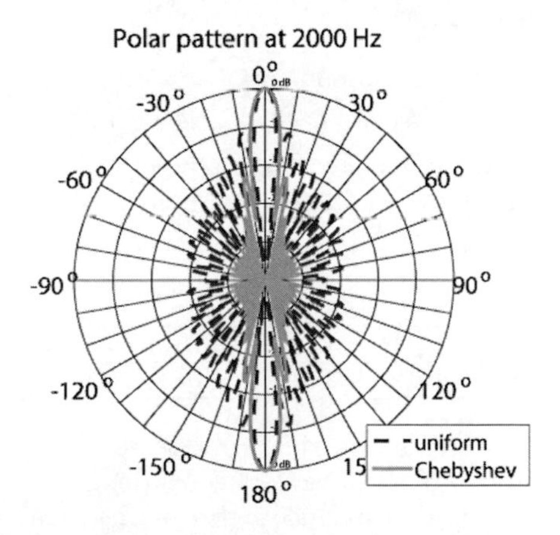

Figure B.1.: Beam patterns of the linear array with uniform and Chebyshev weightings.

The Rayleigh resolution, which is related to BW_{NN} can be calculated by [98]:

$$R(\theta) = 1.22\frac{L}{D}\lambda\frac{1}{cos^3(\theta)} \tag{B.1}$$

where L is the distance from the array center to near-side surface of the train, D is the array diameter, θ is the steering angle between the incident wave direction and the main response axis (MRA) of the array, and λ is the wavelength. The resolution indicates that if two sound sources are within the range, they cannot be distinguished with each other. For instance, the resolution of this array at 2 kHz, 3.2 m distance, with no steering is 0.36 m (6°, steering angle equals zero).

As stated before, the weighting w_n is important to array performance. If uniform weighting is used, the resolution can reach the best when the noise at each microphone is assumed as spatially uncorrelated. Therefore the resolution and the sidelobe level reduction are reciprocal with each other. If the Chebyshev weighting is applied, the MSL increases by 9 dB to 18 dB and at the same time half BW_{NN} increases from 3° to 9° (0.5 m, steering angle equals zero) at 2 kHz compared to the uniform weighting. Taking Chebyshev weighting as an example of the nonuniform weightings is because it increases the beamwidth with the least value and the MSL with only around 1 dB less than the others.

B.2. Measurement setup

Several parameters of the measurements are given in Tab. B.1. Two types of regional trains, RE9 and RB20, in North Rhine-Westphalia, Germany were measured at the speeds of 150 km/h and 91 km/h. A video camera was used to synchronize the position of the trains and the recordings for later processing.

N	D (m)	d (m)	L (m)	H (m)
24	1.84	0.08	3.2	2.64

Table B.1.: Several parameters in the array setup. N is the number of microphones, D is the array diameter (length), d is the microphone spacing, L is the distance between the array and the near-side surface of the train and H is the height of the array (from the top microphone to the ground).

The setup of the measurement is sketched in Fig. B.2. The on-site measurement is shown in Fig. B.3.

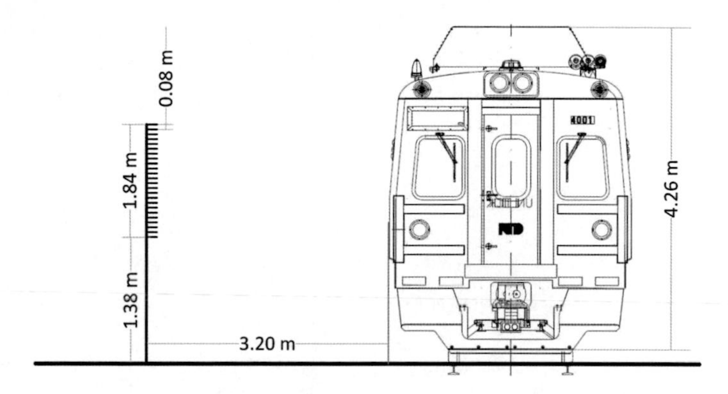

Figure B.2.: Front view of the on-site train pass-by measurement setup.

Figure B.3.: The setup of the train pass-by measurements.

B.3. Localization results

The recorded signals at the microphones are extracted according to different grid points and then processed by DSB. Fig. B.4 shows the processing on a line of vertical grid points. This procedure repeats to calculate all the grid points on the plane (This procedure is the same as in Fig. 4.2).

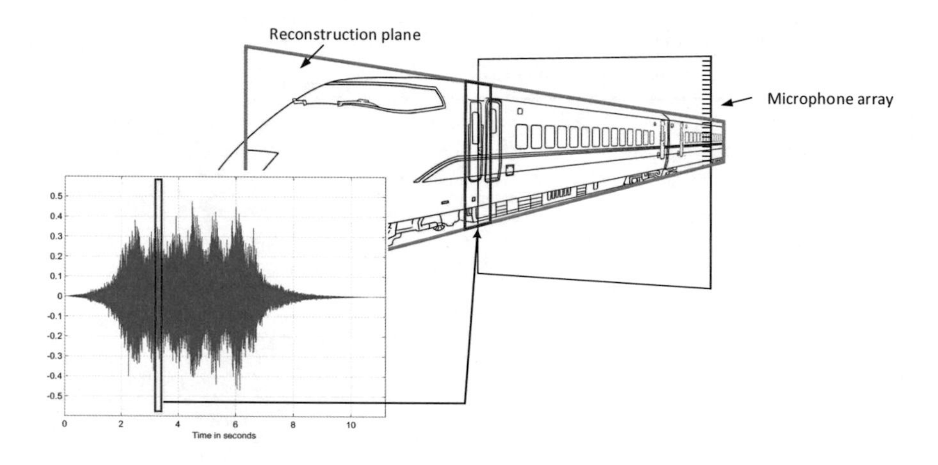

Figure B.4.: The sketch of the processing procedure. The red square on the reconstruction plane consists of a line of vertical grid points to be reconstructed. The red square on the audio signal represents the extracted signal of a microphone in terms of a grid point in the red square on the reconstruction plane.

Uniform weighting is applied at the frequencies below 2.5 kHz, while for higher frequencies Chebyshev weighting is used. It assures higher resolution at lower frequencies and larger restriction of MSL at higher frequencies. Fig. B.5 presents the localization results of RE9 at different 1/3 octave bands. It can be seen that the rolling noise generated by the wheel rail contact is one of the main sound sources. As the frequency increases, the aerodynamic noise gradually increases, mainly from the pantograph, the gaps between coaches and the facilities outside of the train above the roof. Nevertheless, the rolling noise is dominant in a wide frequency range.

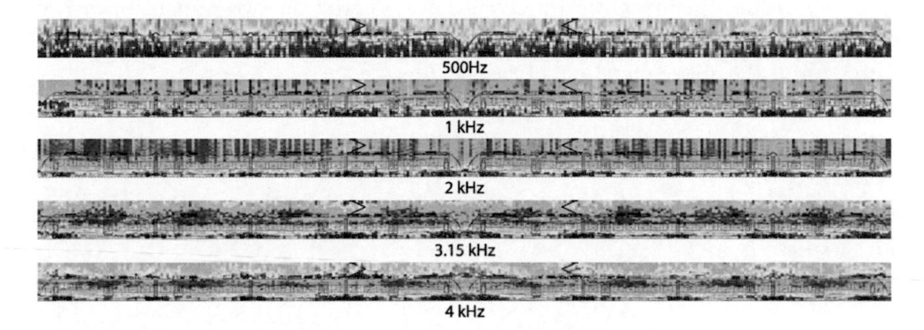

Figure B.5.: The localization results of the RE9 train of in 1/3 octave bands.

Fig. B.6 shows the localization results of RB20 at 1/3 octave bands. The distribution of the noise sources is similar as RE9. Differently, fewer outer facilities lead to less noise contribution from them.

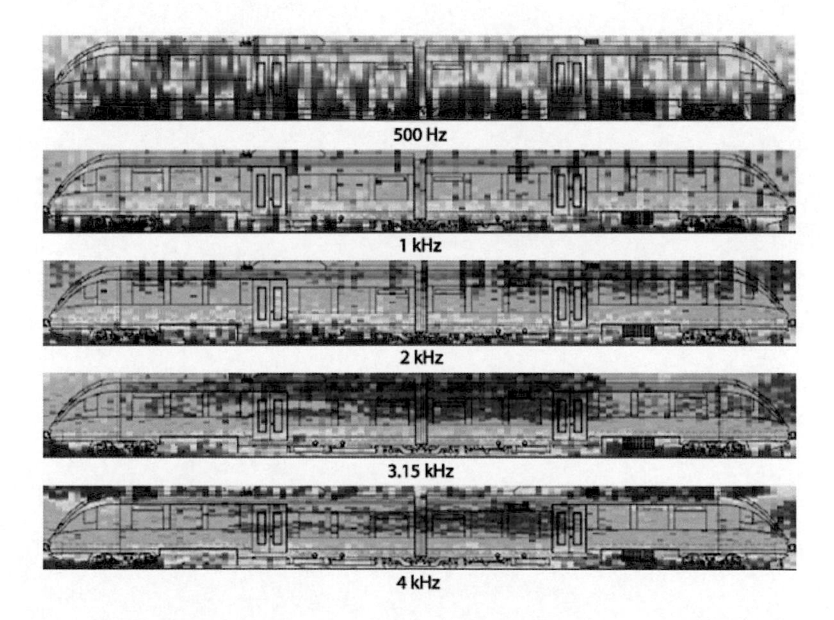

Figure B.6.: The localization results of the RB20 train in 1/3 octave bands.

The resolution of the linear array is limited at low frequencies, e.g. 500 Hz in the two cases above. In addition, the vertical array has no horizontal resolution which disables the array to track the moving sound source to eliminate the Doppler effect. It results in inaccurate localization. Therefore, two-dimension arrays and de-Dopplerization are necessary to address the localization of moving sound sources, and the subsequent signal reconstruction.

C

Separable arrays

To accelerate the beamforming and deconvolution [99] process using kronecker array transform (KAT), separable array geometry is necessary. To achieve comparable results, the resolutions of the spiral and separable arrays should be similar, and the microphone numbers remain the same. Therefore, the diameter of the separable array is set to 0.3 m, half of the size of the spiral array. Non-redundant array [72] is able to keep higher resolution capability using a small number of microphones compared to a larger uniform array. In this research, 6 microphones are aligned linearly, with the microphone spacing set to 0.02 m, 0.07 m, 0.12 m, 0.26 m and 0.3 m. Extending the linear non-redundant array to two-dimensional, a 6×6 array is obtained. After eliminating microphones in the four corners of the array aperture, the reduced number of the microphones remains identical with the spiral array. The configuration is shown in Fig. C.1. Fig. C.2 provides the beam patterns of the spiral, the 6 × 6 separable and the reduced 6 × 6 separable arrays at 3 kHz. First of all, after removing the microphones in the corners, the beam pattern has no significant change. Second, the spiral and reduced separable arrays share similar beam width, and so is the resolution. The sidelobe levels of the separable arrays are almost 10 dB higher than that of spiral array. Nevertheless, the sidelobes are not relevant by applying appropriate deconvolution methods since the sidelobes can be significantly constrained.

A plane (1.5 m × 5 m) moves in the x-direction, together with two point sources at the speed of 40 m/s. Two point sources are placed on the plane with 2 m spacing. The microphone array is set at 1.5 m away from the moving direction. The array origin is on the z-axis. The plane is meshed into grids, with 5 cm spacing between each other. Each grid represents a potential sound source, so that the array can steer its angle to "scan" the plane to search for the sources. The left source consists of a 2 kHz tone and noise, and the right source signal contains the same noise as in the left one. In both cases, additive white Gaussian noise (AWGN) with SNR = 20 dB is added. The sound pressure RMS of the

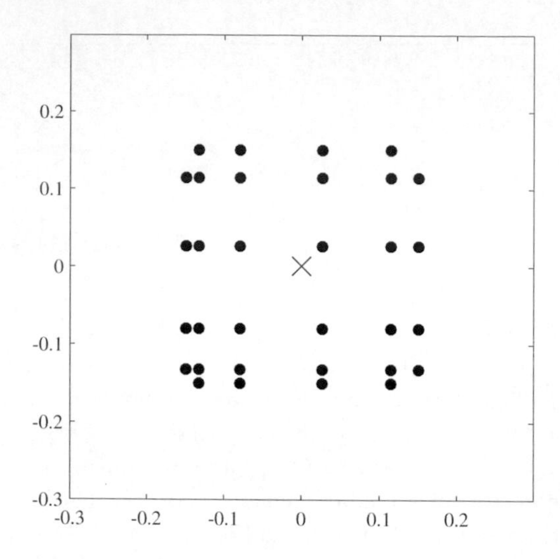

Figure C.1.: The separable array. The "●" represents the position of a microphone and the "×" represents the origin of the array. The unit of the axes is meter.

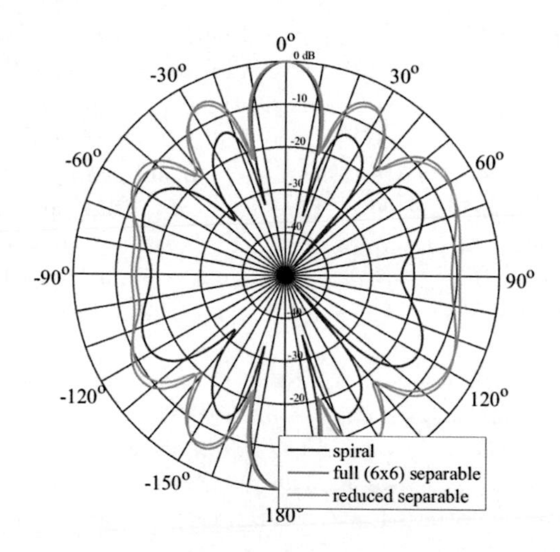

Figure C.2.: The beam patterns of the spiral, the 6 × 6 separable and the reduced 6 × 6 separable arrays.

tone and noise are both 1 Pa. The good localization ability of the deconvolution method using spiral and separable arrays has been confirmed to be comparable [13]. Therefore, only the color map generated by DSB id shown.

Fig. C.3 shows the localization results using the spiral and separable arrays. As what mentioned before, the separable array delivers larger slidelobe levels. Both arrays are able to localize the left periodic source at 2 kHz. With the arrays

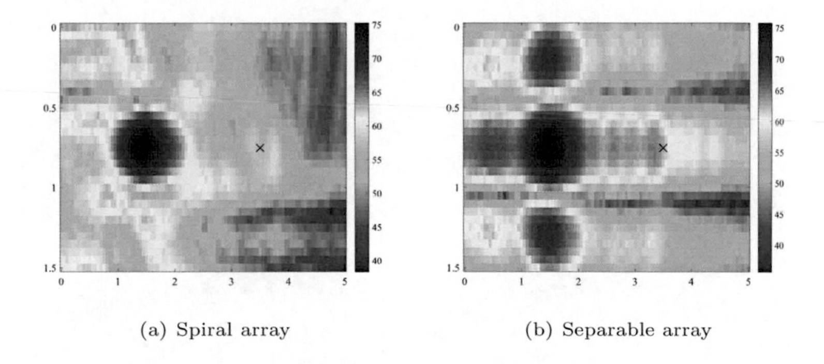

(a) Spiral array (b) Separable array

Figure C.3.: The localization results of the spiral and separable arrays in the 1/3 octave band with 2 kHz center frequency. The × represents the source position.

steering to the left source, the periodic signal is reconstructed (Fig. C.4). The similar spectra indicate that with accurate localization results, the two arrays deliver similar signal reconstruction.

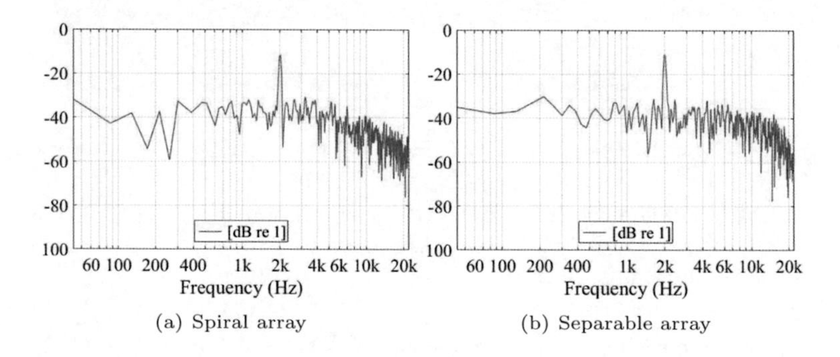

(a) Spiral array (b) Separable array

Figure C.4.: The signal reconstruction results of the spiral and separable arrays in the 1/3 octave band with 2 kHz center frequency.

Acronyms

AWGN	additive white Gaussian noise
CB	compressive beamforming
CS	compressive sensing
CWT	continuous wavelet transform
CroPaC	cross pattern coherence
DSB	delay and sum beamforming
FFT	fast Fourier transform
FT	Fourier transform
ITA	Institute of Technical Acoustics
LCMV	linear constrained minimum variance
JNDF	Just-noticeable difference in frequency
JNDL	Just-noticeable difference in level
KAT	kronecker array transform
MPDR	minimum power distortionless response
MRA	main response axis
MUSIC	multiple signal classification
MVDR	minimum variance distortionless response
MSL	maximum sidelobe level
RIP	restricted isometry property
RMS	root mean square
SMS	spectral modeling synthesis
SNR	signal-to-noise ratio
SPL	sound pressure level
SR	sparse recover
STFT	short-time Fourier transform
StRIP	statistical restricted isometry property
TDTF	time-domain transfer function
VA	Virtual Acoustics
VR	Virtual reality
WT	wavelet transform

Bibliography

[1] WHO Regional Office for Europe. Burden of disease from environmental noise - quantification of healthy life years lost in europe. 2011.

[2] Maarten Hornikx. Ten questions concerning computational urban acoustics. *Building and Environment*, 106:409–421, 2016.

[3] Jian Kang. *Urban sound environment.* CRC Press, 2006.

[4] Michael Vorlaender. *Auralization: Fundamentals of Acoustics, Modelling, Simulation, Algorithms and Acoustic Virtual Reality.* Springer Berlin Heidelberg, Berlin, Heidelberg, 2007.

[5] Noam R. Shabtai, Gottfried Behler, Michael Vorländer, and Stefan Weinzierl. Generation and analysis of an acoustic radiation pattern database for forty-one musical instruments. *The Journal of the Acoustical Society of America*, 141(2):1246, 2017.

[6] Fanyu Meng, Gottfried Behler, and Michael Vorländer. A synthesis model for a moving sound source based on beamforming. *Acta Acustica united with Acustica*, 104(2):351–362, 2018.

[7] Stephen A. Rizzi and Brenda M. Sullivan. Synthesis of virtual environments for aircraft community noise impact studies. In *Proceedings of the 11th AIAA/CEAS Aeroacoustics Conference (26th AIAA Aeroacoustics Conference)*, volume 4, 23-25 May 2005.

[8] Ho-Chul Shin, Cesare Hall, and Daniel Crichton. Auralisation of turbofan engine noise components. In *Proceedings of the 12th AIAA/CEAS Aeroacoustics Conference (27th AIAA Aeroacoustics Conference)*, [Reston, VA], 5-9 June 2006.

[9] Mathieu Sarrazin, Karl Janssens, Herman van der Auweraer, Wim Desmet, and Paul Sas. Virtuel car sound synthesis approach for hybrid and electric vehicles. In *SAE International 2012*, June 2012.

[10] Abhishek Kumar Sahai, Eckhard Anton, Eike Stumpf, Frank Wefers, and Michael Vorländer. Interdisciplinary auralization of take-off and landing procedures for subjective assessment in virtual reality environments. In *Proceedings of the 18th AIAA/CEAS Aeroacoustics Conference (33rd AIAA Aeroacoustics Conference)*, Reston, Virigina, 04 - 06 June 2012.

[11] M. Arntzen and D. G. Simons. Modeling and synthesis of aircraft flyover noise. *Applied Acoustics*, 84:99–106, 2014.

[12] Abhishek Kumar Sahai. *Consideration of Aircraft Noise Annoyance during Conceptual Aircraft Design.* Phd dissertation, RWTH Aachen University, Aachen, Germany, 2016.

[13] Abhishek Sahai, Frank Wefers, Sebastian Pick, Eike Stumpf, Michael Vorländer, and Torsten Kuhlen. Interactive simulation of aircraft noise in aural and visual virtual environments. *Applied Acoustics*, 101:24–38, 2016.

[14] Abhishek K. Sahai, Mirjam Snellen, and Dick G. Simons. Objective quantification of perceived differences between measured and synthesized aircraft sounds. *Aerospace Science and Technology*, 72:25–35, 2018.

[15] Reto Pieren, Kurt Heutschi, Menno Müller, Madeleine Manyoky, and Kurt Eggenschwiler. Auralization of wind turbine noise: Emission synthesis. *Acta Acustica united with Acustica*, 100(1):25–33, 2014.

[16] Reto Pieren, Thomas Bütler, and Kurt Heutschi. Auralization of accelerating passenger cars using spectral modeling synthesis. *Applied Sciences*, 6(1):5, 2015.

[17] Jens Forssén, ALICE HOFFMANN, and Wolfgang Kropp. Auralization model for the perceptual evaluation of tyre–road noise. *Applied Acoustics*, 132:232–240, 2018.

[18] Reto Pieren, Kurt Heutschi, Jean Marc Wunderli, Mirjam Snellen, and Dick G. Simons. Auralization of railway noise: Emission synthesis of rolling and impact noise. *Applied Acoustics*, 127:34–45, 2017.

[19] Like Jiang, Massimiliano Masullo, Luigi Maffei, Fanyu Meng, and Michael Vorländer. A demonstrator tool of web-based virtual reality for participatory evaluation of urban sound environment. *Landscape and Urban Planning*, 2017.

[20] Jan Jagla, Julien Maillard, and Nadine Martin. Sample-based engine noise synthesis using an enhanced pitch-synchronous overlap-and-add method. *The Journal of the Acoustical Society of America*, 132(5):3098–3108, 2012.

[21] Martin Klemenz. Sound synthesis of starting electric railbound vehicles and the influence of consonance on sound quality. *Acta Acustica united with Acustica*, 91(4):779–788, 2005.

[22] FrederikEmpa Rietdijk, KurtEmpa Heutschi, and Christoph Zellmann. Determining an empirical emission model for the auralization of jet aircraft: Zenodo. In *Proceedings of the 10th European Congress and Exposition on Noise Control Engineering*, 31 May -3 June 2015.

[23] Frederik Rietdijk. *Auralisation of airplanes considering sound propagation in a turbulent atmosphere*. Phd dissertation, Chalmers University of Technology, Gothenburg, Sweden, 2017.

[24] Andrew Peplow, Jens Forssén, Peter Lundén, and Mats E. Nilsson. Exterior auralization of traffic noise within the listen project. In *Proceedings of the European Conference on Acoustics (Forum Acusticum 2011)*, 23 June - 1 July 2011.

[25] Estelle Bongini, Stéphane Molla, Pierre-Etienne Gautier, Dominique Habault, Pierre-Olivier Mattéi, and F. Poisson. Synthesis of noise of operating vehicles: development within silence of a tool with listening features. In *Noise and Vibration Mitigation for Rail Transportation Systems: Proceedings of the 9th International Workshop on Railway Noise*, pages 320–326, Munich, Germany, 4-8 September 2007.

[26] B. D. van Veen and K. M. Buckley. Beamforming: a versatile approach to spatial filtering. *ASSP Magazine, IEEE*, 5(2):4–24, 1988.

[27] Harry L. van Trees. *Optimum Array Processing: Detection, estimation, and modulation theory*. John Wiley & Sons, Inc, New York, USA, 2002.

[28] Don H. Johnson and Dan E. Dudgeon. *Array signal processing: concepts and techniques*. Simon & Schuster, 1992.

[29] B. Barsikow, W.F. King, and E. Pfizenmaier. Wheel/rail noise generated by a high-speed train investigated with a line array of microphones. *Journal of Sound and Vibration*, 118(1):99–122, 1987.

[30] S. Brühl and A. Röder. Acoustic noise source modelling based on microphone array measurements. *Journal of Sound and Vibration*, 231(3):611–617, 2000.

[31] Vincent Fleury and Jean Bulté. Extension of deconvolution algorithms for the mapping of moving acoustic sources. *The Journal of the Acoustical Society of America*, 129(3):1417–1428, 2011.

[32] Diange Yang, Ziteng Wang, Bing Li, Yugong Luo, and Xiaomin Lian. Quantitative measurement of pass-by noise radiated by vehicles running at high speeds. *Journal of Sound and Vibration*, 330(7):1352–1364, 2011.

[33] Florent Le, Jean-Hugh Thomas, Franck Poisson, and Jean-Claude Pascal. Genetic optimisation of a plane array geometry for beamforming. application to source localisation in a high speed train. *Journal of Sound and Vibration*, 371:78–93, 2016.

[34] G. P. Howell, A. J. Bradley, M. A. McCormick, and J. D. Brown. Dedopplerization and acoustic imaging of aircraft flyover noise measurements. *Journal of Sound and Vibration*, 105(1):151–167, 1986.

[35] H. Kook, G. B. Moebs, P. Davies, and J. S. Bolton. An efficient procedure for visualizing the sound field radiated by vehicles during standardized passby tests. *Journal of Sound and Vibration*, 233(1):137–156, 2000.

[36] Hugo Elias Camargo. *A Frequency Domain Beamforming Method to Locate Moving Sound Sources.* Phd dissertation, Virginia Polytechnic Institute and State University, Virginia, USA, 2010.

[37] Pieter Sijtsma, Stefan Oerlemans, and Hermann Holthusen. Location of rotating sources by phased array measurements. In *Proceedings of 7th AIAA/CEAS Aeroacoustics Conference and Exhibit*, Reston, Virigina, 2001.

[38] J. Hald J.J. Christensen. Technical review: Beamforming, 2004.

[39] Jack Capon. High-resolution frequency-wavenumber spectrum analysis. *Proceedings of the IEEE*, 57(8):1408–1418, 1969.

[40] Ralph O. Schmidt. Multiple emitter location and signal parameter estimation. *IEEE Transactions on Antennas and Propagation*, 34(3):276–280, 1986.

[41] T. Noohi, N. Epain, and C. T. Jin. Super-resolution acoustic imaging using sparse recovery with spatial priming: 2015 ieee international conference on acoustics, speech and signal processing (icassp). In *Proceedings of IEEE International Conference on Acoustics, Speech and Signal Processing (ICASSP 2015)*, Piscataway, NJ, 2015. IEEE.

[42] N. Epain and C. T. Jin. Super-resolution sound field imaging with sub-space pre-processing. In *Proceedings of 2013 IEEE International Conference on Acoustics, Speech and Signal Processing (ICASSP)*, pages 350–354, Piscataway, NJ, May 26-31 2013. IEEE.

[43] McCormack, L., Delkaris-Manias, S. and Pulkki, V. Parametric acoustic camera for real-time sound capture, analysis and tracking. In *Proceedings of the 20th Internation Conference of Digital Audio Effects (DAFx-17)*, 2017.

[44] Dmitry Malioutov, Müjdat Cetin, and Alan S. Willsky. A sparse signal reconstruction perspective for source localization with sensor arrays. *IEEE Transactions on signal processing*, 53(8):3010–3022, 2005.

[45] Ali Cafer Gurbuz, James H. McClellan, and Volkan Cevher. A compressive beamforming method. In *Acoustics, Speech and Signal Processing, 2008. ICASSP 2008. IEEE International Conference on*, pages 2617–2620. IEEE, 2008.

[46] Geoffrey F. Edelmann and Charles F. Gaumond. Beamforming using compressive sensing. *The Journal of the Acoustical Society of America*, 130(4):EL232–EL237, 2011.

[47] Angeliki Xenaki, Peter Gerstoft, and Klaus Mosegaard. Compressive beamforming. *The Journal of the Acoustical Society of America*, 136(1):260–271, 2014.

[48] Peter Gerstoft, Angeliki Xenaki, and Christoph F. Mecklenbräuker. Multiple and single snapshot compressive beamforming. *The Journal of the Acoustical Society of America*, 138(4):2003–2014, 2015.

[49] Heinrich Kuttruff. *Acoustics: An introduction.* Taylor & Francis, London and New York, english ed. edition, 2007.

[50] Philip M. Morse and K. Uno Ingard. *Theoretical acoustics.* Princeton University Press, Princeton, 1986(1968).

[51] B. Barsikow and W. F. King. On removing the doppler frequency shift from array measurements of railway noise. *Journal of Sound and Vibration*, 120(1):190–196, 1988.

[52] Ann P. Dowling and J. E. Ffowcs Williams. *Sound and sources of sound.* Horwood, 1983.

[53] Emmanuel J. Candè and Michael B. Wakin. An introduction to compressive sampling. *Signal Processing Magazine, IEEE*, 25(2):21–30, 2008.

[54] Emmanuel J. Candes. The restricted isometry property and its implications for compressed sensing. *Comptes Rendus Mathematique*, 346(9-10):589–592, 2008.

[55] Raghu S. Raghunathan, H.-D Kim, and T. Setoguchi. Aerodynamics of high-speed railway train. *Progress in Aerospace Sciences*, 38(6-7):469–514, 2002.

[56] David Thompson. *Railway noise and vibration: mechanisms, modelling and means of control*. Elsevier, 2008.

[57] Stylianos Kephalopoulos, Marco Paviotti, and Fabienne Anfosso Ledee. Common noise assessment methods in europe (cnossos-eu). *Common noise assessment methods in Europe (CNOSSOS-EU)*, pages 180–p, 2012.

[58] Florent Le Courtois and Julien Bonnel. Compressed sensing for wideband wavenumber tracking in dispersive shallow water. *The Journal of the Acoustical Society of America*, 138(2):575–583, 2015.

[59] Dmitry Malioutov. *A sparse signal reconstruction perspective for source localization with sensor arrays*. Phd dissertation, Massachusetts Institute of Technology, Cambridge, Massachusetts, US, 2003.

[60] Michael Grant and Stephen Boyd. Graph implementations for nonsmooth convex programs. *Recent advances in learning and control*, pages 95–110, 2008.

[61] D. Gabor. Theory of communication. part 1: The analysis of information. *Journal of the Institution of Electrical Engineers - Part III: Radio and Communication Engineering*, 93(26):429–441, 1946.

[62] Fanyu Meng and Michael Vorländer. Synthesis of moving sound sources using wavelet transform. In *Proceedings of the 46th International Congress and Exposition on Noise Control Engineering (INTER-NOISE 2017)*, 27-30 August 2017.

[63] Xavier Serra. *A system for sound analysis/transformation/synthesis based on a deterministic plus stochastic decomposition*. Phd dissertation, Stanford University, California, USA, 1989.

[64] Xavier Serra and Julius Smith. Spectral modeling synthesis: A sound analysis/synthesis system based on a deterministic plus stochastic decomposition. *Computer Music Journal*, pages 12–24, 1990.

[65] Chinmay Pendharkar. *Auralization of road vehicles using spectral modeling synthesis.* Master thesis, Chalmers University of Technology, Gothenburg, Sweden, 2012.

[66] Eberhard Zwicker and Hugo Fastl. *Psychoacoustics: Facts and models.* Springer Science & Business Media, 2013.

[67] Lawrence R. Rabiner and B. H. Juang. *Fundamentals of speech recognition.* Prentice Hall signal processing series. PTR Prentice Hall and London : Prentice-Hall International (UK), Englewood Cliffs, NJ, 1993.

[68] Markus M-Trapet. *Comparison of Sound-Source Localization Methods for Vibrating Structures.* Master thesis, RWTH Aachen University, Aachen, Germany, 2009.

[69] Bernard Ginn, Jesper Gomes, and Jørgen Hald. Recent advances in rail vehicle moving source beamforming. In *Proceedings of the 42nd International Congress and Exposition on Noise Control Engineering 2013 (INTER-NOISE 2013)*, Red Hook, NY, 2013. Curran.

[70] Fanyu Meng, Frank Wefers, and Michael Vorlaender. Acquisition of exterior multiple sound sources for train auralization based on beamforming. In *Proceedings of the 10th European Congress and Exposition on Noise Control Engineering*, pages 1703–1708, 31 May -3 June 2015.

[71] Mingsian R. Bai, Jeong-Guon Ih, and Jacob Benesty. *Acoustic Array Systems: Theory, Implementation, and Application.* John Wiley & Sons, 2013.

[72] E. Vertatschitsch and S. Haykin. Nonredundant arrays. *Proceedings of the IEEE*, 74(1):217, 1986.

[73] Dane Bush and Ning Xiang. Broadband implementation of coprime linear microphone arrays for direction of arrival estimation. *The Journal of the Acoustical Society of America*, 138(1):447–456, 2015.

[74] Hugh C. Pumphrey. Design of sparse arrays in one, two, and three dimensions. *The Journal of the Acoustical Society of America*, 93(3):1620–1628, 1993.

[75] Flávio P. Ribeiro and Vítor H. Nascimento. Fast transforms for acoustic imaging—part i: Theory. *Image Processing, IEEE Transactions on*, 20(8):2229–2240, 2011.

[76] Flávio P. Ribeiro and Vítor H. Nascimento. Fast transforms for acoustic imaging–part ii: Applications. *IEEE transactions on image processing : a publication of the IEEE Signal Processing Society*, 20(8):2241–2247, 2011.

[77] Fanyu Meng and Michael Vorländer. Using microphone arrays to reconstruct moving sound sources for auralization. In *Proceeedings of European Acoustics Association (EAA) Euroregio*, 13-15 June 2016.

[78] Bruno Masiero and Vitor H. Nascimento. Revisiting the kronecker array transform. *IEEE Signal Processing Letters*, 24(5):525–529, 2017.

[79] Thomas F. Brooks and William M. Humphreys. A deconvolution approach for the mapping of acoustic sources (damas) determined from phased microphone arrays. *Journal of Sound and Vibration*, 294(4):856–879, 2006.

[80] Fanyu Meng, Bruno Masiero, and Michael Vorländer. Compressive beamforming for moving sound source auralization. In *Proceedings of the 45th International Congress and Exposition on Noise Control Engineering (INTER-NOISE 2016)*, 21-24 August 2016.

[81] H. Kook, P. Davies, and J. S. Bolton. Statistical properties of random sparse arrays. *Journal of Sound and Vibration*, 255(5):819–848, 2002.

[82] Charles F. Gaumond and Geoffrey F. Edelmann. Sparse array design using statistical restricted isometry property. *The Journal of the Acoustical Society of America*, 134(2):EL191–7, 2013.

[83] Peter Gerstoft and William S. Hodgkiss. Improving beampatterns of two-dimensional random arrays using convex optimization. *The Journal of the Acoustical Society of America*, 129(4):EL135–40, 2011.

[84] Sifa Zheng, Feng Xu, Xiaomin Lian, Yugong Luo, Diange Yang, and Keqiang Li. Generation method for a two-dimensional random array for locating noise sources on moving vehicles. *Noise Control Engineering Journal*, 56(2):130–140, 2008.

[85] Mingsian R. Bai, You Siang Chen, and Yi-Yang Lo. A two-stage noise source identification technique based on a farfield random parametric array. *The Journal of the Acoustical Society of America*, 141(5):2978, 2017.

[86] ISO 10844-1994. Acoustics. specification of test tracks for the purpose of measuring noise emitted by road vehicles, 15 March 1995.

[87] Emmanuel Candes and Terence Tao. The dantzig selector: Statistical estimation when p is much larger than n. *The Annals of Statistics*, pages 2313–2351, 2007.

[88] Yuejie Chi, Louis L. Scharf, Ali Pezeshki, and A. Robert Calderbank. Sensitivity to basis mismatch in compressed sensing. *IEEE Transactions on Signal Processing*, 59(5):2182–2195, 2011.

[89] Marco Berzborn, Ramona Bomhardt, Johannes Klein, Jan-Gerrit Richter, and Michael Vorländer. The ita-toolbox: An open source matlab toolbox for acoustic measurements and signal processing. In *Proceedings of the 43th Annual German Congress on Acoustics (DAGA)*, 6-9 March 2017.

[90] José A. Ballesteros, Ennes Sarradj, Marcos D. Fernández, Thomas Geyer, and Maria Jesús Ballesteros. Methodology for pass-by measurements on cars with beamforming. In *Proceedings of the 5th Berlin Beamforming Conference (BeBeC)*, Berlin, 2014.

[91] Tapio Lokki. *Physically-based auralization: design, implementation, and evaluation.* PhD thesis, Helsinki University of Technology, Helsinki, Finland, 2002.

[92] Lauri Savioja, Jyri Huopaniemi, Tapio Lokki, and Ritta Väänänen. Creating interactive virtual acoustic environments. *Journal of the Audio Engineering Society*, 47(9):675–705, 1999.

[93] H. G. Jonasson. Acoustical source modelling of road vehicles. *Acta Acustica united with Acustica*, 93(2):173–184, 2007.

[94] Xuetao Zhang. *Applicable directivity description of railway noise sources.* PhD thesis, Chalmers University of Technology, Gothenburg, Sweden, 2011.

[95] Institute of Technical Acoustics, RWTH Aachen University. Virtual acoustics - a real-time auralization framework for scientific research, 2018.

[96] Tobias Lentz. *Binaural technology for virtual reality: Lehrstuhl und Institut für Technische Akustik.* Phd dissertation, Logos-Verl and Zugl.: Aachen, Techn. Hochsch., Diss., 2007, 2008.

[97] Dirk Schröder. *Physically based real-time auralization of interactive virtual environments.* Phd dissertation, RWTH Aachen University, Aachen, Germany, 2012.

[98] David Havelock, Sonoko Kuwano, and Michael Vorländer. *Handbook of signal processing in acoustics.* Springer Science & Business Media, 2008.

[99] Robert P. Dougherty. Extensions of damas and benefits and limitations of deconvolution in beamforming. In *Proceedings of the 11th AIAA/CEAS Aeroacoustics Conference (26th AIAA Aeroacoustics Conference)*, 23-25 May 2005.

Curriculum Vitæ

Personal Data

Name	Fanyu Meng
Date of birth	29.05.1988
Birth place	Kedong Heilongjiang, China

Educational Background

08/1995–07/2000	Kedong Shiyan Primary School, China
08/2000–6/2004	Kedong No. Two Middle School, China
08/2004–06/2007	Kedong No. One Middle School, China

Higher Education

08/2007–07/2011	Bachelor's degree in Industrial Design at Harbin Institute of Technology, China
08/2011–07/2013	Master's degree in Mechanical Design and Theory at Harbin Institute of Technology, China

Professional Experience

10/2013–9/2017	Research Assistant at the Institute of Technical Acoustics, RWTH Aachen University, Germany
10/2017–present	Researcher at Eindhoven University of Technology, Netherlands

August 7, 2018

Bisher erschienene Bände der Reihe

Aachener Beiträge zur Akustik

ISSN 1866-3052
ISSN 2512-6008 (seit Band 28)

1	Malte Kob	Physical Modeling of the Singing Voice ISBN 978-3-89722-997-6 40.50 EUR
2	Martin Klemenz	Die Geräuschqualität bei der Anfahrt elektrischer Schienenfahrzeuge ISBN 978-3-8325-0852-4 40.50 EUR
3	Rainer Thaden	Auralisation in Building Acoustics ISBN 978-3-8325-0895-1 40.50 EUR
4	Michael Makarski	Tools for the Professional Development of Horn Loudspeakers ISBN 978-3-8325-1280-4 40.50 EUR
5	Janina Fels	From Children to Adults: How Binaural Cues and Ear Canal Impedances Grow ISBN 978-3-8325-1855-4 40.50 EUR
6	Tobias Lentz	Binaural Technology for Virtual Reality ISBN 978-3-8325-1935-3 40.50 EUR
7	Christoph Kling	Investigations into damping in building acoustics by use of downscaled models ISBN 978-3-8325-1985-8 37.00 EUR
8	Joao Henrique Diniz Guimaraes	Modelling the dynamic interactions of rolling bearings ISBN 978-3-8325-2010-6 36.50 EUR
9	Andreas Franck	Finite-Elemente-Methoden, Lösungsalgorithmen und Werkzeuge für die akustische Simulationstechnik ISBN 978-3-8325-2313-8 35.50 EUR

10	Sebastian Fingerhuth	Tonalness and consonance of technical sounds	ISBN 978-3-8325-2536-1	42.00 EUR
11	Dirk Schröder	Physically Based Real-Time Auralization of Interactive Virtual Environments	ISBN 978-3-8325-2458-6	35.00 EUR
12	Marc Aretz	Combined Wave And Ray Based Room Acoustic Simulations Of Small Rooms	ISBN 978-3-8325-3242-0	37.00 EUR
13	Bruno Sanches Masiero	Individualized Binaural Technology. Measurement, Equalization and Subjective Evaluation	ISBN 978-3-8325-3274-1	36.50 EUR
14	Roman Scharrer	Acoustic Field Analysis in Small Microphone Arrays	ISBN 978-3-8325-3453-0	35.00 EUR
15	Matthias Lievens	Structure-borne Sound Sources in Buildings	ISBN 978-3-8325-3464-6	33.00 EUR
16	Pascal Dietrich	Uncertainties in Acoustical Transfer Functions. Modeling, Measurement and Derivation of Parameters for Airborne and Structure-borne Sound	ISBN 978-3-8325-3551-3	37.00 EUR
17	Elena Shabalina	The Propagation of Low Frequency Sound through an Audience	ISBN 978-3-8325-3608-4	37.50 EUR
18	Xun Wang	Model Based Signal Enhancement for Impulse Response Measurement	ISBN 978-3-8325-3630-5	34.50 EUR
19	Stefan Feistel	Modeling the Radiation of Modern Sound Reinforcement Systems in High Resolution	ISBN 978-3-8325-3710-4	37.00 EUR
20	Frank Wefers	Partitioned convolution algorithms for real-time auralization	ISBN 978-3-8325-3943-6	44.50 EUR

21	Renzo Vitale	Perceptual Aspects Of Sound Scattering In Concert Halls ISBN 978-3-8325-3992-4 34.50 EUR
22	Martin Pollow	Directivity Patterns for Room Acoustical Measurements and Simulations ISBN 978-3-8325-4090-6 41.00 EUR
23	Markus Müller-Trapet	Measurement of Surface Reflection Properties. Concepts and Uncertainties ISBN 978-3-8325-4120-0 41.00 EUR
24	Martin Guski	Influences of external error sources on measurements of room acoustic parameters ISBN 978-3-8325-4146-0 46.00 EUR
25	Clemens Nau	Beamforming in modalen Schallfeldern von Fahrzeuginnenräumen ISBN 978-3-8325-4370-9 47.50 EUR
26	Samira Mohamady	Uncertainties of Transfer Path Analysis and Sound Design for Electrical Drives ISBN 978-3-8325-4431-7 45.00 EUR
27	Bernd Philippen	Transfer path analysis based on in-situ measurements for automotive applications ISBN 978-3-8325-4435-5 50.50 EUR
28	Ramona Bomhardt	Anthropometric Individualization of Head-Related Transfer Functions Analysis and Modeling ISBN 978-3-8325-4543-7 35.00 EUR
29	Fanyu Meng	Modeling of Moving Sound Sources Based on Array Measurements ISBN 978-3-8325 4759-2 44.00 EUR